YOUR BRAIN
IS A TIME
MACHINE

WITHDRAWN

ALSO BY DEAN BUONOMANO

*Brain Bugs: How the
Brain's Flaws Shape Our Lives*

W. W. NORTON & COMPANY
Independent Publishers Since 1923
NEW YORK • LONDON

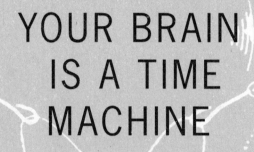

YOUR BRAIN IS A TIME MACHINE

The Neuroscience and
Physics of Time

Dean Buonomano

For information about permission to reproduce selections from this book,
write to Permissions, W. W. Norton & Company, Inc.,
500 Fifth Avenue, New York, NY 10110

For information about special discounts for bulk purchases, please contact W. W. Norton Special
Sales at specialsales@wwnorton.com or 800-233-4830

Manufacturing by Quad Graphics, Fairfield
Book design by Ellen Cipriano
Production manager: Louise Mattarelliano

Library of Congress Cataloging-in-Publication Data

Names: Buonomano, Dean.
Title: Your brain is a time machine : the neuroscience and physics of time /
   Dean Buonomano.
Description: First edition. | New York : W. W. Norton & Company, [2017] |
   Includes bibliographical references and index.
Identifiers: LCCN 2016046898 | ISBN 9780393247947 (hardcover)
Subjects: LCSH: Brain—Physiology. | Time perception.
Classification: LCC QP376 .B8635 2017 | DDC 612.8/2—dc23 LC record
   available at https://lccn.loc.gov/2016046898

W. W. Norton & Company, Inc.
500 Fifth Avenue, New York, N.Y. 10110
www.wwnorton.com

W. W. Norton & Company Ltd.
15 Carlisle Street, London W1D 3BS

1 2 3 4 5 6 7 8 9 0

TO ANA

CONTENTS

PART I: BRAIN TIME

1:00	Flavors of Time	3
2:00	The Best Time Machine You'll Ever Own	17
3:00	Day and Night	34
4:00	The Sixth Sense	57
5:00	Patterns in Time	79
6:00	Time, Neural Dynamics, and Chaos	101

PART II: THE PHYSICAL AND MENTAL NATURE OF TIME

7:00	Keeping Time	127
8:00	Time: What the Hell Is It?	144
9:00	The Spatialization of Time in Physics	157
10:00	The Spatialization of Time in Neuroscience	179
11:00	Mental Time Travel	195
12:00	Consciousness: Binding the Past and the Future	215

ACKNOWLEDGMENTS	235
NOTES	237
BIBLIOGRAPHY	255
INDEX	279

PART I

Brain Time

1:00 FLAVORS OF TIME

All that really belongs to us is time; even he
who has nothing else has that.

—BALTASAR GRACIÁN

time

person

year

way

day

What do the words in the above list have in common?

One would certainly be forgiven for not recognizing them as the five most commonly used nouns in the English language.[1] That the word *time* sits atop the list, along with two others that are units of time, is a consequence of the overwhelming importance time plays in our lives. When we are not asking for the time, we are speaking of *saving time, killing time, serving time, keeping time, not having time, tracking time, bedtime, time outs, buying time, good times, time travel, overtime, free time,* and my personal favorite, *lunchtime.*

For their part, scientists and philosophers talk about *subjective time, objective time, proper time, coordinate time, sidereal time, emergent time, time perception, encoding time, relativistic time, time cells, time dilation*, *reaction time*, *spacetime*, and the rather redundant *Zeit-geber* (time giver) *time*.

Ironically, although *time* is the most common noun, there is no consensus on how it should be defined. Indeed, the inherent challenge in attempting to define time was famously captured over 1,600 years ago by the Christian philosopher Saint Augustine: "What then is time? If no one asks me, I know what it is. If I wish to explain it to him who asks, I do not know."

Few questions are as perplexing and profound as those that relate to time. Philosophers ponder what time is, and whether it is a single moment or a full-blown dimension. Physicists grapple with why time appears to flow in only one direction, whether time travel is possible, and even whether time exists at all. Neuroscientists and psychologists, in turn, struggle to understand what it means to "feel" the passage of time, how the brain tells time, and why humans are uniquely capable of mentally projecting ourselves into the future. And time is at the heart of the question of free will: is the future an open path, or preordained by the past?

The goal of this book is to explore and, to the extent possible, answer these questions. As we begin, however, we must first acknowledge that our ability to answer questions pertaining to time is constrained by the nature of the organ asking them. Although the gelatinous mass of 100 billion brain cells stashed within your skull is the most sophisticated device in the known universe, it was not "designed" to understand the nature of time any more than your laptop was designed to write its own software. Thus, as we explore questions of time, we will learn that our intuitions and theories about time reveal as much about the nature of time as they do about the architecture and limitations of our brains.

THE DISCOVERY OF TIME

Time is complicated, more so than *space*.

Yes, it is true that space has more dimensions than time: three values are necessary to pinpoint a location in space (for example, latitude, longitude, and elevation), whereas only one number is needed to mark a moment in time. So in some sense *space* is more complicated, but what I mean is that it is much harder for the human brain to understand *time* than *space*.

Consider our fellow vertebrates, with whom we share much of our neural hardware. Vertebrate animals are able to navigate through space, create an internal map of their surroundings, and in a sense, "understand" the concept of space. Animals migrate over large distances with a clear objective as to where in space they are heading; they remember where they stored their food; and even a puppy knows that if a treat falls behind the couch, she can try to circumnavigate the couch and access the treat from the left, right, below, or above. We know that the brains of mammals have a highly sophisticated internal map of space because neuroscientists have been recording from so-called *place cells* in the hippocampus for over four decades. Place cells are neurons that fire, or "turn on," when an animal is located in a specific place in a room—that is, a particular point in space. Together, these cells form a network that creates a spatial map of the external world somewhat like a GPS system, except much more flexible; for example, our internal spatial maps seem to be instantly updated when the boundaries of a room are altered or objects are moved about.

Animals can not only navigate through space, they can "see" it.[2] Standing upon a mountain, we can see the sky above, the forest below, and a winding river flowing into the ocean—each in its place in space. We can also "hear" space—that is, locate the point in space from where a sound is coming. Our sense of touch (somatosensation) informs us of not only the position and shape of objects, but of the location in space of our most important possessions: our limbs.

Time is different. Animals, of course, cannot physically navigate through time. Time is a road without any bifurcations, intersections, exits, or turnarounds. Perhaps for this reason, there was relatively little evolutionary pressure for animals to map, represent, and understand time with the same fluency as space. We will see that animals certainly tell time and anticipate when events will take place, but it seems unlikely that our vertebrate relatives can be said to understand the differences between past, present, and future in the same way that their brain grasps the differences between up, down, left, and right. Our sensory organs do not directly detect the passage of time.[3] Unlike the fictional Tralfamadorians of Kurt Vonnegut's novel *Slaughterhouse-Five*, we cannot see across time, taking in the past, present, and future at a single glance.

The brains of all animals, humans included, come better equipped to navigate, sense, represent, and understand space than time. Indeed one of the theories of how humans came to understand the concept of time is that the brain co-opted the circuits already in place to represent and understand space (chapter 10). As we will see, this may be one reason all cultures seem to use spatial metaphors to talk about time (it was a *long* day, I'm looking *forward* to the eclipse, in *hindsight* I should not have said that out loud).

Time is more complicated than space for scientists as well. Fields of science, like humans, undergo developmental stages: they mature and change as they grow. And in many fields one signature of this maturation process is the progressive embrace of time.

The first true field of modern science was arguably geometry, formalized by Euclid in the third century BC. Geometry is often defined as "the branch of mathematics concerned with the properties and relations of points, lines, surfaces, and solids."[4] Euclidean geometry is notable both because it is one of the most elegant and transformative theories in the history of science, and because it achieved this stature despite its total disregard of time. Geometry could have just as well been called *spaceometry*: the study of things that are frozen in time and never change. Geometry was one of the first true scientific fields for a reason: science is much simpler if one can get away with ignoring time.

The mathematics available to the Greek philosophers and scientists was not well suited to studying how things change over time. Furthermore, during antiquity it was much easier to measure distance than time; today the opposite is true, as we can measure time much more precisely than space (chapter 7). It took close to two thousand years after Euclid to begin to fully incorporate time into math and physics. An important step in this direction took place in the late sixteenth century when, according to a probably apocryphal story, a bored Galileo Galilei noted that the time it took a swaying lamp in the Cathedral of Pisa to complete a full cycle was independent of the amplitude of the swing—that is, the period of the swing was the same whether it was a broad or narrow swing (it was later determined that the period does increase slightly with amplitude).[5] By studying motion, how the position of objects changes over time, Galileo helped give birth to the field of dynamics. But like the Greeks, Galileo also lacked the tools to mathematically define the relationships between forces, motion, speed, and acceleration. It took Newton and Leibniz to invent the ultimate mathematical tool to capture how things change over time: *calculus*.[6] Using calculus, Newton was able to describe the laws that govern the motion of falling apples and orbiting planets alike.

Newton believed in absolute time, one that "from its own nature flows equably without regard to anything external." For him there was a true and universal time that applied unequivocally to all points in space. Newton's universe appeared to be a deterministic one: all of time, past and future, could in principle be determined from the present alone. But there were many further scientific advances in store. Two are particularly relevant to us. First, little by little scientists reached the disheartening (to some) realization that even in a universe that fully obeyed Newton's beautiful laws, it was not possible *in practice* to predict the future (or retrodict the past). The work of many scientists, including the French mathematician Henri Poincaré and the American meteorologist Edward Lorenz, revealed that tiny differences in the state of a system can lead to vastly different future outcomes (the most famous example being the *butterfly effect* in weather prediction). This is called *chaos*, and we will see that it rears its head when we study the most complex dynamical system we know of—our brains (chapter 6). The second advance was that Albert Einstein swept away Newton's notion of absolute and universal time. Against all intuition, Einstein established that time was relative (chapter 9). We will discuss this topic in detail, but for now the point is that as the field of physics matured, the problem of time became progressively engrained and fundamental. Up to a point. Ironically, in some corners there is a push to eliminate time altogether from physics,[7] returning us to a static geometrical universe, which the physicist Julian Barbour refers to as *Platonia*—an allusion to Plato's notion that ideal geometric forms are real entities that exist in a timeless realm.

TIME AND NEUROSCIENCE

Many other fields of science also underwent a similar maturation process. For example, modern biology began in the eighteenth century as a fairly descriptive and static taxonomy of life forms, but biology grew to incorporate time in the form of evolution and dynamics. Darwin played the role of Galileo: he saw that species on planet Earth were in constant "motion": mutating, disappearing, and evolving.

The fields of neuroscience and psychology also evolved to progressively incorporate the problem of time. Say what you will about the pseudoscience of phrenology, at least phrenologists acknowledged the significance of our sense of time. They assigned our sense of time to an area of the frontal lobes conveniently located in between *tune* and *space* ("locality") (Figure 1.1). According to one phrenology text, "The office of this faculty is to mark the passage of time, duration, succession of events, etc. It also remembers dates, keeps correct time in music and dancing, and induces to punctuality in the fulfillment of engagements."[8]

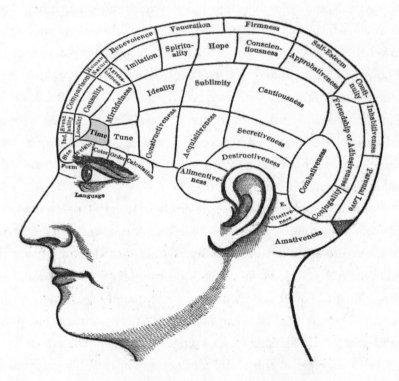

Figure 1.1: A phrenology chart from the nineteenth century.

William James, one of the fathers of modern psychology, also recognized the importance of time to understanding the mind. Indeed, he devoted a chapter of his magnum opus, *The Principles of Psychology* (published in 1890), to the perception of time. Oddly, since then few landmark books in psychology or neuroscience have done the same.[9] Indeed, throughout most of the twentieth century, the problem of time was somewhat neglected and largely omitted from textbooks.

I am oversimplifying a bit. First, the problem of time in neuroscience and psychology is not a single problem, but a set of interconnected problems relating to how the brain tells time, generates complex temporal patterns, consciously perceives the passage of time, recollects the past, and thinks about the future. Second, significant progress was made in many subfields relating to the psychology and neuroscience of time. For example, the field of *chronobiology*, the study of biological rhythms, most notably sleep-wake cycles, flourished throughout the twentieth century (chapter 3). Additionally, throughout this same period there were many pioneers who advanced our understanding of how the brain tells and perceives time. But relatively speaking, the problems pertaining to time have been overlooked. Pick up the bible of modern neuroscience, the textbook *Principles of Neural Science*, and look for the most common noun in the English language in the index—you will not find it. Whereas, if you look up "space," you will find it represented in multiple entries.[10]

Psychology and neuroscience are newborn scientific fields, they are only beginning to fully grasp the importance of time and dynamics. As the University of California, Berkeley, psychologist Richard Ivry wrote in 2008, "A generation ago, research on timing was limited, emphasizing the study of behaviors marked by temporal regularities. More recently, a renaissance has taken hold in the study of time perception, with researchers addressing a broad range of temporal phenomena."[11]

As an example of this shift, consider one of the holy grail questions of psychology and neuroscience: *how does the brain store memories?* As memories pertain to past experiences, memory is inherently entwined with time. But even here, scientists have often neglected to place the problem of memory in its correct temporal context. It is only in the twenty-first century that scientists are beginning to fully embrace the fact that "information about the past is useful only to the extent that it allows us to anticipate what may happen in the future."[12] Memory did not evolve to allow us to reminisce about the past. The sole evolutionary function of memory is to allow animals to predict what will happen, when it will happen, and how to best respond when it does happen. Thanks to ongoing conceptual shifts, together with a multitude of methodological advances, there has been increasing focus on time in neuroscience and psychology. And, critically, there is increasing recognition that without an understanding of how the brain tells time, perceives time, and represents time, it will not be possible to understand the human mind.

PRESENTISM VERSUS ETERNALISM

This book focuses primarily on the neuroscience and psychology of time, but we will also delve into questions pertaining to the physics of time. Here the goal is not only to understand some of the fundamental insights physics provides about the nature of time, but also to explore where the neuroscience and physics of time intersect—or perhaps I should say where they *clash* (chapters 8 and 9). Toward this goal it will be important to introduce the two most important philosophical theories on the nature of time: *presentism* and *eternalism*.

Presentism, as the name hints, states that only the present is real. Under presentism, the past is a configuration of the universe that once existed, and the future refers to some yet-to-be-determined configuration. Eternalism, in sharp contrast, states that the past and future are as equally real as the present. There is absolutely nothing particularly special about the present: under eternalism *now* is to time as *here* is to space. Even though you currently find yourself to be in one point in space, you know that there are many other points in space—different rooms, cities, planets, and galaxies—that are all equally valid places to be in. Similarly, even though you perceive yourself to be in a point in time you call *now*, there are past and future moments in time in which other beings, and younger and older *yous*, find themselves (Figure 1.2).

Perhaps the simplest way to grasp the distinction between presentism and eternalism is in the context of the topic of time travel.[13] Under presentism, true time travel (jumping back and forth between

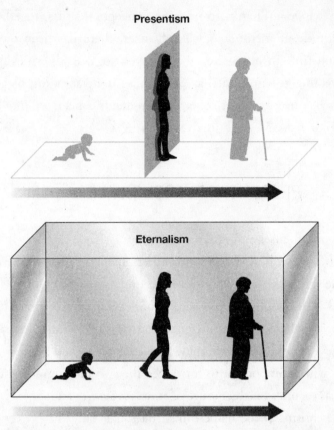

Figure 1.2: Two views of the nature of time: presentism versus eternalism.

the past and future) is a nonstarter. Technical considerations, such as whether it is possible to build a time machine or whether the laws of physics allow it, are irrelevant; one simply cannot travel to a time that does not exist any more than one can travel to a place that does not exist. Under eternalism, time is a dimension much (but not exactly) like space, so the universe is a four-dimensional "block"—one in which the past and future are as real as locations north and south of you. Although eternalism is agnostic as to whether time travel is achievable, it validates the discussion because there would be "places" (moments) in time to travel to.

Presentism certainly conforms to our intuition that as each instant of our lives transforms into a past moment, it is gone. Whether or not that moment leaves an imprint in our memory, the moment itself ceases to exist. Presentism also corroborates our feeling of control: that our decisions and actions shape an open future. Neuroscientists rarely have to grapple with the issue of presentism versus eternalism. But in practice, neuroscientists are implicitly presentists. They view the past, present, and future as fundamentally distinct, as the brain makes decisions in the present, based on memories of the past, to enhance our well being in the future. But despite its intuitive appeal, presentism is the underdog theory in physics and philosophy.

Versions of eternalism go back at least two-and-a-half millennia, to the Greek philosopher Parmenides, who believed we live in a timeless world in which there is no change. Today, for very good reasons, most physicists and philosophers accept the eternalist stance that all of time is in some sense "already" laid out within the block universe. It is not that the notion of time as a fourth dimension is simply a convenient mathematical abstraction—like representing time along the x-axis of a graph—but rather that the past, present, and future truly stand on equal footing.

Now for the clash between neuroscience and physics: if all moments in time are equally real, and all events in our past and future are eternally embedded within the block universe, then our perception of the *flow of time* must be an illusion (chapter 9). In other words, if all of time is already "out there," then time is not *flowing* or *passing* in the normal sense of those words. As the philosopher Jack Smart once put it, "Talk of the flow of time or the advance of consciousness is a dangerous metaphor that must not be taken literally."[14] So it would seem that one of the most unequivocal and universally shared subjective experiences—that time is passing—must be relegated to some sort of trick of the conscious mind. This is indeed a widely held view. For example, in his book *Time's Arrow and Archimedes' Point* the philosopher Huw Price writes: "Philosophers have tended to divide into two camps on these issues. On the one side are those who treat flow and the present as objective features of the world [presentism]; on the other, those who argue that these things are mere artifacts of our subjective perspective on the world [eternalism]. . . . I shall be taking the latter view for granted."

The mathematician and physicist Herman Weyl famously captured the clash between our perception of time and the standard block universe view when he stated: "The objective world simply is, it does not happen. Only to the gaze of my consciousness, crawling upward along the world line of my body, does a section of the world come to life as a fleeting image in space which continuously changes in time."[15]

THE PLURAL OF TIME

Any discussion of time, whether in neuroscience, philosophy, or physics, is inevitably muddled by the fact that the word *time* is used to mean many different things. One of the reasons it is the most used noun in the English language is because it is actually multiple words. Indeed, the different uses of the word *time* vary from language to language. In English we say *speed is distance divided by time*, and ask *what time is it*? Portuguese has two different words for these contexts. The word *tempo* is used to define speed, but to find out the time one would ask, *que horas são* (what hour is it). But in contrast to English, in Portuguese one would use the word *tempo* to inquire about the *weather*.

We seamlessly use the different meanings of the word *time* in our day-to-day life, but such seamlessness inevitably clouds attempts to rigorously explore questions relating to time. So it will be helpful to, if not define, at least constrain, some of the different meanings of the word. Consider the following sentence: "Minkowski's talk on the nature of time ended on time, but it seemed to drag on for a long time."

This contrived sentence attempts to capture three meanings of the word *time* that will be important for our goals. In order, I will refer to them as *natural time*, *clock time*, and *subjective time*.

Intuitively we understand time to be the medium in which our lives play out. I use the term *natural time* (as in "the nature of time") to refer to the concept of time as this medium or "dimension." Natural time is the flavor of time at the core of the presentism-versus-eternalism debate. In practice most scientists can ignore questions pertaining to natural time, but ultimately what could be more profound than knowing whether other "versions" of ourselves are laid out along the temporal dimension of the block universe, or than determining whether our sense of the passage of time is merely one of many illusions the brain bestows upon the mind.

For practical purposes, *time* is sometimes defined as *what clocks tell*. As circular as it may seem, this definition is an extremely important one. But it does, inevitably, lead to the question: *what exactly is a clock?* In the most general sense, a clock is a device that undergoes changes in some reproducible manner, and offers a way to quantify these changes. The changes could be represented in the swings of a pendulum, the vibrations of a quartz crystal, or even the amount of radioisotopes of carbon in a fossil sample. *Clock time* is the most used sense of the word *time* in science. Einstein, however, stressed that "Such a definition is satisfactory when we are concerned with defining a time exclusively for the place where the watch is located; but it is no longer satisfactory when we have to connect in time, series of events occurring at different places."[16] *Clock time* is a local measure of change, neither absolute nor universal. Nevertheless, clock time is ultimately what rules our lives: it not only tells us when to rise, work, and sleep, but, because the body itself is a clock, clock time governs when we grow old and expire.

Subjective time refers to our conscious sense of time: the subjective feeling of both the passage of time, and of *how much time* has passed. Like all subjective experiences, subjective time is a construct created by the brain—it does not exist outside the confines of the skull. Much as our subjective perception of color allows us to experience a physical property of visible light (wavelength), our subjective sense of time is a mental construct that allows us to, in a sense, "feel" both *natural time* and *clock time*.

Philosophers and scientists have pondered the mysteries of time for millennia. And yet, one thousand and six hundred years after Saint Augustine vented about the challenge of defining time, we still don't know the answer to questions as fundamental as whether the past, present, and future are equally real, or whether our perception of the passage of time is an illusion.

Before fully answering such questions, the field of neuroscience will have to further mature and embrace the fact that it will not be possible to understand the human mind without describing how the brain tells, represents, and conceptualizes time. This is because, as I argue in the next chapter, the brain is a time machine: a machine that not only tells time and predicts the future, but one that allows us to mentally project ourselves forward in time. It is exceedingly easy to overlook the fact that without the ability to mentally travel into the future, our species would have never crafted an obsidian stone into a tool, or grasped that by planting seeds today we can ensure our future survival.

Our unique ability to grasp the concept of time and peer into the distant future is, however, both a gift and curse. Over the course of evolution we went from being subjugated to nature's unpredictable and capricious ways to overruling Mother Nature herself: manipulating the present to assure survival in the future. But our clairvoyant abilities also led to the inevitable realization that our own time is finite and fleeting. Gift or curse, we are now faced with a wonderful and perplexing mystery: *What is time?*

2:00 THE BEST TIME MACHINE YOU'LL EVER OWN

Any real body must have extension in four directions:
it must have Length, Breadth, Thickness, and—Duration.
But through natural infirmity of the flesh, which I will explain
to you in a moment, we incline to overlook this fact. There are
really four dimensions, three which we call the three planes
of Space, and a fourth, Time.

—H. G. WELLS, 1895

Hollywood has ensured that we are all familiar with the concept of time travel. *The Terminator, Groundhog Day, Back to the Future, The Time Traveler's Wife, Looper, Midnight in Paris, Interstellar,* and some high percentage of the *Star Trek* films represent a small sample of movies that have exposed us to the mind-bending paradoxes that arise from hopping backward and forward through time—such as traveling back in time and accidentally committing grandpacide.

Despite its current ubiquity in movies, books, and TV, and even as a serious topic of study in physics, the concept of time travel is conspicuously absent from most of human history. The Bible, along with other religious texts and orally transmitted folktales, are full of stories of talking animals, gods, and other supernatural beings. They tell of animals transmuting into humans and vice versa, epic voyages over vast spatial distances, Methuselah-like humans whose lives have spanned centuries, magic, and resurrections. But, oddly enough, little or no time travel. Even Shakespeare, who seems to have anticipated the plots and twists of almost every modern movie, never touched upon the subject of time travel. There are exceptions of sorts; for example, the *Mahabharata*, an ancient Hindu poem from around 800 BC, tells the story of a king and his daughter who visit the god Brahma to seek out a worthy groom. They later learn that during the time of their visit, many generations have passed on Earth, along with the king's possessions and treasures. So there are Rip Van Winkle-ish, relativistic stories of time passing at different rates, but no jumping back and forth between moments of time. Charles Dickens's *A Christmas Carol*, written in the mid-nineteenth century, was a precursor to time-travel stories. In it, Ebenezer Scrooge is led to Christmases past and future by ghosts, but the voyage is a dreamlike, passive one—there is no interaction between characters from different points in time. It was only in the late nineteenth century that the notion of true time travel emerged, most famously in H. G. Wells's *The Time Machine*, in which the protagonist travels to the future, interacts with the atrophied descendants of humankind, and returns to his present time.[1]

Why was true time travel absent from fiction until the end of the nineteenth century? Perhaps because human beings are innate presentists: few things are as obvious as the fact that the past is irrevocably gone and thus immutable, and that the future does not yet exist. Perhaps the notion that the past and the future are as real as the present, and thus potential travel destinations, was simply too counterintuitive and fantastical to be incorporated even into fiction. So what changed in the late nineteenth century that opened the gates of time travel in our imaginations? It is difficult to answer this question, but certainly a scientific revolution was brewing.

A key event in this revolution culminated with the publication of

Einstein's theory of special relativity in 1905, which forever shattered our intuitions of the nature of time. Einstein established that clocks would tick at different rates depending on the speed at which they were traveling. Two years after that, Einstein's mathematics professor, Hermann Minkowski, demonstrated that, mathematically speaking, Einstein's theory could be elegantly placed in the framework of a 4D universe—that is, a universe in which time was literally another dimension, much like space.

We will explore the physics that led to the acceptance of the 4D block universe in chapter 9, but for now the point is that in the twentieth century, little by little, time travel became an acceptable topic of study in physics. Not so much because most physicists believed that true time travel into the past or future was actually possible, but because no one was able to prove that it was not. Many physicists accept that in principle there are "places" in time to travel to, but nevertheless believe that for practical or theoretical reasons, the laws of physics will prohibit jumping back and forth between them.[2] This is because time travel has rather exotic requirements. Wormholes are perhaps the least implausible mode of time travel. Imagine the surface of the Earth as a sheet of space and time, and then building a tunnel as a shortcut between Washington and Beijing. While compliant with the current laws of physics, wormholes are hypothetical entities. Time travel would require not only creating or finding one, but moving one of its openings around at very high speeds. And then hoping that the wormhole is stable and traversable, meaning that whoever goes in would not be—to use the scientific term—spaghettified.

But I digress, for my goal in this chapter is not to discuss the plausibility or implausibility of true time travel, but to convince you that your brain is the best time machine you will ever own. Or put in another way, you are the best time machine that has ever been built.

THE BRAIN IS A TIME MACHINE

Of course the brain does not allow us to physically travel through time, but it is a time machine of sorts for four interrelated reasons:

1. **The brain is a machine that remembers the past in order to predict the future.** Over hundreds of millions of years, animals have engaged in a race to predict the future. Animals foresee the actions of prey, predators, and mates; they prepare for the future by caching food and building nests; and they anticipate dawn and dusk, spring and winter. The degree to which animals succeed in divining the future translates into the evolutionary currency of survival and reproduction. Consequently, the brain is at its core a prediction or anticipation machine.[3] And whether you realize it or not, on a moment-by-moment basis your brain is automatically attempting to predict what is about to _____. These short-term predictions, up to a few seconds into the future, are entirely automatic and unconscious. If a bouncy ball rolls off the table, we automatically adjust our movements to catch it off the bounce, which we do not do when a muffin falls off the table.

    Humans and other animals are also continuously attempting to make long-term predictions. The simple act of an animal surveying its environment is an attempt to peer into the minutes and hours that lie ahead: as a wolf stops to take in the sights, sounds, and odors around it, it is searching for clues that will help it avoid potential predators and find prey and mates. In order to predict the future the brain stores a vast amount of information about the past; and like Apple's backup software Time Machine, it sometimes adds temporal labels (dates) to these memories, allowing us to review episodes of our lives organized on a timeline.

2. **The brain is a machine that tells time.** Your brain performs a wide range of computations, including those necessary to recognize a face, or to choose your next move in chess. Telling time is another type of computation the brain performs: not simply measuring the seconds, hours, and days of our lives, but recognizing and generating temporal patterns, such as the intricate rhythms of a song, or the carefully timed sequence of movements that allow a gymnast to perform a round-off backflip.

   Telling time is a critical component of predicting the future. As any meteorologist knows, it is not sufficient to announce that it will rain; one must also predict *when* it will rain. As a cat launches into the air to catch a bird in flight, it must predict where the bird will be a second into the future. Pollinating birds, in turn, are known to keep track of the amount of time that elapsed since their last visit to a particular flower, in order to allow the nectar to be replenished before the next visit.[4] From the ability to throw a spear at a moving target, time the punch line of a joke, or play Beethoven's *Moonlight Sonata* on the piano to the ability to regulate daily sleep-wake cycles and monthly reproductive cycles, virtually every aspect of animal behavior and cognition requires the ability to tell time.

3. **The brain is a machine that creates the sense of time.** Unlike vision or hearing, we do not have a sensory organ that detects time. Time is not a form of energy or a fundamental property of matter that can be detected via physical measurements. Yet, much in the same way that we consciously perceive the color of objects (the wavelengths of reflected electromagnetic radiation), we consciously perceive the passage of time. The brain creates the feeling of the passage of time. Like most subjective experiences, our sense of time undergoes many illusions and

distortions. The same duration—as measured by an external clock—can seem to fly by or drag depending on a multitude of factors. But distorted or not, the conscious perception of the passage of time, and that the world around us is in continuous flux, is among the most familiar and undeniable experiences of all. Yet, it is this feeling of the passage of time that is fundamentally at odds with the notion of time held by many physicists and philosophers.

4. **The brain allows us to mentally travel back and forth in time.** The race to predict the future was won, hands down, by our hominin ancestors when they developed the ability to understand the concept of time and mentally project themselves backward into the past and forward into the future— that is, to engage in *mental time travel* (chapter 11). As Abraham Lincoln reportedly said, "The best way to predict the future is to create it," and this is exactly what mental time travel allowed us to do. We went from predicting nature's capricious ways to creating the future by overruling nature herself.

The influential Canadian psychologist Endel Tulving explained: "Early expressions of future-oriented thought and planning consisted of learning to use, preserve, and then make fire, to make tools, and then to store and carry these with them. Furnishing the dead with grave goods; growing their own crops, fruits, and vegetables; domesticating animals as sources of food and clothing; . . . these all represent relatively recent developments in human evolution. Every single one is predicated on the awareness of the future."[5]

We have all mentally re-experienced the joy or sorrow of past events and run alternate simulations of those episodes to explore what could have been. In the other direction we jump into the future every time we dread or daydream about what may come, and we simulate different plot lines of our future lives in the hope of determining the best course of action in the present. Humans may or may not be the only creatures on the planet to engage in mental time travel, but we are certainly the only animals to use this ability to ponder the possibility of actually traveling to the past or future.

TIME AS A TEACHER

In the eighteenth century, the Scottish philosopher David Hume pondered how we make sense of the world—how we figure out the relationships between events that occur at different points of space and moments in time. He emphasized three principles underlying human understanding: *resemblance* (the similarity between objects and events), *contiguity* (the temporal and spatial "proximity" of events), and *cause and effect*. In regard to cause and effect, he provided a number of rules that we use to determine if two events are causally related to each other, including:

1. The cause and effect must be contiguous in space and time.
2. The cause must be prior to the effect.[6]

Fortunately, one does not need to read Hume to put such rules into effect, because they are hardwired into our brains at the level of synapses and neurons. The temporal relationships between events are among the most important clues the brain uses to make sense of what William James referred to as the "blooming, buzzing, confusion" of sensory information assaulting our sensory organs. How does a baby learn that the word *cat* refers to fluffy, four-legged creatures with sharp claws? Because the first dozen times a baby sees a cat, parents coo, "Look at the kitty kat." In other words, the temporal contiguity between the sight of the cat and the sound of the word *cat* is what allows a baby's neural circuits to link those two distinct stimuli to each other.

One of the most universal forms of learning in the animal kingdom, classical conditioning, captures the fundamental importance of temporal contiguity and order to brain function. Pavlov's dog is the standard example of classical conditioning: ring a bell (the conditioned stimulus) before presenting meat (the unconditioned stimulus) to a dog, and he will eventually learn to salivate in response to the sound of the bell alone—or, perhaps more familiarly, your cat may learn to materialize in the kitchen at the sound of the can opener. Although the bell doesn't actually *cause* the food, it might as well, as far as the dog is concerned. Classical conditioning is the primordial algorithm animals use to predict what is about to happen next. Rattlesnakes are slithering, living examples of a classical conditioning experiment: the rattle serves as the conditioned stimulus that a rattlesnake (the unconditioned stimulus) may be underfoot.

Hume could have never suspected just how important temporal contiguity is to brain function. Consider the challenge faced by the brain of a baby in recognizing her mother's face. The face sometimes appears up close, and thus large, but at other times is seen from the distance, and is thus small. At each distance the image projected onto the retina is totally different (that is, just like a close-up or distance picture of the same person will activate different patterns of pixels on a camera, different spatial patterns of photoreceptors on the retina will be activated). So how does a baby learn that all those distinct images correspond to Mom? This so-called *size invariance* problem is a complex one, and it is not known how the brain solves it. But one theory is that it uses temporal contiguity. Part of a baby's experience is seeing Mom grow and shrink as she either approaches or walks away. If the brain assumes that the vastly distinct patterns impinging on the retina that occur in temporal contiguity are from the same object, it could eventually learn the general principles of size invariance: those patterns that occurred consecutively represent the same object in the external world.[7] Put another way, the prediction is that if temporal contiguity were removed—imagine growing up in stroboscopic world in which snapshots of objects appeared to magically jump from small to big to small every ten seconds—the ability to recognize an object viewed at different distances as one and the same would be impaired.

Classical conditioning, and many other forms of learning, capture the essence of the temporal asymmetry mentioned in Hume's second rule: that cause must precede effect. When Pavlov presented the meat before the sound of the bell, no conditioning occurred. Similarly, classical conditioning is highly sensitive to the degree of temporal contiguity, more specifically the interval between the events. If Pavlov presented the meat one hour after ringing the bell, the relationship between the bell and food becomes absolutely invisible to the dog—even though the bell still predicts the appearance of food. Indeed, for the most part animals seem to be incapable of connecting the temporal dots between events separated by minutes or hours, much less days or months.[8] The longer the interval between two events, the harder it is to detect the connection. Classical conditioning is a shortsighted form of learning.

It takes more complex cognitive abilities to understand the relationships between events separated by days, months, and years. Our ability to conceptualize time and engage in mental time travel is what allows us to see the relationship between sex and childbirth, or seeds and trees. But we too are temporally myopic: if cigarettes caused cancer a week after starting to smoke, as opposed to many decades later, the tobacco industry would never have become a worldwide trillion-dollar industry (chapter 11).

TEMPORAL DIRECTION AND MISDIRECTION

It is impossible to overstate the importance to cognition of the temporal relationship between the events we experience. For example, as cognitive psychologist Steven Pinker has pointed out, we generally assume that the order in which events are stated reflects the order in which they occurred. Thus the one-liner: *they got married and had a baby—but not necessarily in that order.* In most languages it is easier to understand the relationship between events that are stated in the order they occurred:[9] *she smiled before opening the gift* is easier to process than *before opening the gift, she smiled.*

The brain's assumptions about temporal order and interval allow us to understand and anticipate the events unfolding in the world, but these assumptions can also be misleading. Consider a trick in which a magician picks up a coin from a table with her right hand, then flamboyantly bumps her clenched hands together while reciting "Abracadabra," and finally reveals that the coin is no longer in either hand. The trick relies on temporal misdirection.[10] The disappearance of the coin is automatically assumed to be caused by the "most contiguous" event: the bump and exaggerated *abracadabra.* In reality, of course, the coin was never in either hand, as the trick lies in the sleight of hand of sliding the coin off the table as the magician goes through the motions of picking it up. Once again, the longer the interval between two events, the harder it is to see the relationship between them. By inserting a gap between the true cause of the disappearance of the coin and the reveal, magicians exploit our built-in temporal assumptions.

In my previous book, *Brain Bugs*, I described an example of temporal misdirection encountered while playing blackjack for the first time in Las Vegas. I knew that blackjack consists of hoping that the two cards you are dealt add up to 21, and if they do not, deciding whether you should take another card and run the risk of "busting" (overshooting 21). The dealer is your opponent, and he plays like an automaton, taking an additional card until his sum adds up to 17 or more. I figured that if I played by the same strategy as the dealer, my odds of winning any given hand should be 50-50. Of course, I knew the house always has the advantage, but I could not see where it was. It turns out that the house's advantage is straightforward: if both the dealer and

I "bust," he wins. But why could I not see this? The house's advantage is actually hidden from us by temporal misdirection. Here's how it works: since I play first, the dealer immediately gathers my cards and chips when I bust, making it abundantly clear that the game is over for me. He then proceeds to finish the round with the other patrons before revealing his cards. At which point, if I'm still at the table, I might find out that the dealer also busted—and thus that we *should have tied*. I couldn't see the house's advantage because it was hidden in the future: the normal cause-and-effect relationship had been temporally reversed. In a sense, during hands in which we both busted, the effect of my loss comes before the cause: my cards and chips are removed (the effect), before I know whether I lost to or tied with the dealer. It was difficult to see the house's advantage because I ceased to look for it after I was already out of the game. By exploiting this temporal blind spot, casinos hide how the rules are rigged in their favor.[11]

SYNAPTIC CAUSE AND EFFECT

Whether or not we live in the fixed block universe of eternalism, where the passage of time turns out to be illusory, the order of events and the interval between them sculpts our neural circuits. The rules outlined by Hume are in effect algorithms that govern the brain's wiring diagram. The temporal asymmetry of cause and effect, for example, is codified at the most fundamental level within the brain.

Your brain is composed of a network of close to 100 billion neurons, communicating with each other through hundreds of trillions of synapses.[12] Like most computational elements, including the transistors of a computer, neurons receive inputs and generate outputs (Figure 2.1). Compared to transistors, however, neurons are extroverts. The transistor on an average computer chip is connected to maybe a few dozen others; in contrast, the average neuron is connected to thousands of others. These connections are implemented by the syn-

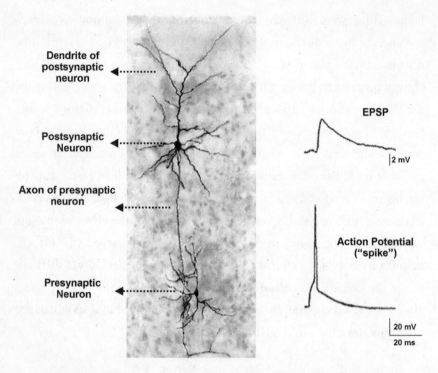

Figure 2.1: Neurons and synapses. Image of two cortical neurons. The axon of the lower, presynaptic neuron connects to a dendrite of the upper, postsynaptic neuron via a synapse (not visible). An action potential—a fast "spike" in the voltage—in the presynaptic neuron produces a small increase in the voltage of the postsynaptic neuron (called an excitatory postsynaptic potential, EPSP). (Modified with permission from Feldmeyer et al., 2002)

apses, the interface between two neurons: a *presynaptic* neuron that is sending a signal out and a *postsynaptic* neuron that is receiving the signal. The inputs to any given neuron come from its presynaptic partners, each providing bioelectrical whispers. Excitatory synapses encourage the postsynaptic neuron to "fire"—that is, generate an output by sending an electrical signal to all its downstream neurons (its own postsynaptic partners). In contrast, inhibitory synapses attempt to persuade the postsynaptic neuron to keep quiet. With so many neurons the nervous system is the wiring diagram from hell. What determines which neurons are connected to which?

For an oversimplified analogy we can look at the World Wide

Web, which is also a network of interconnected elements. Think of the webpages as the neurons, and their unidirectional links as the synapses. Which pages are linked to each other is, for the most part, imposed by outside forces: human code writers. But the brain must wire itself; there is no master programmer. Furthermore, unlike the Web, for the brain it is not only a question of which elements should be connected to each other, but of what the strength of each connection should be. The strength of a synapse refers to the degree to which a presynaptic neuron influences the behavior of the postsynaptic neuron: a strong excitatory synapse between neurons A and B, means that the firing of A is likely to lead to the firing of B, whereas a very weak synapse between neurons A and B means that B doesn't really give a damn about what A is telling it to do. Which neurons are connected to which, and the strength of the synapse between them, is determined in part by synaptic algorithms—so called *synaptic learning rules*—programmed into our genes. So our genes do not encode the strength of the synapses, but they determine the algorithms that govern the strength of the synapses.[13]

One learning rule in particular, *spike-timing-dependent plasticity* (STDP), beautifully illustrates how the temporal asymmetry of cause and effect is built into our synapses. Consider the two neurons shown in Figure 2.2: neuron A is connected to B, and B in turn to A. Thus there are two synapses: $A{\rightarrow}B$ and $B{\rightarrow}A$. We would say these neurons are *recurrently connected*: neuron A is the input to neuron B, and vice versa. Now let's assume that each neuron is driven by distinct events in the outside world. Perhaps the owner of these two neurons is a baby named Zoe, and neuron A is driven by the sound of the letter z, and neuron B by the sound of the letter o; thus, whenever Mom and Dad say Zoe's name, neuron A will fire right before neuron B and, for argument's sake, let's say that neuron A consistently fires 25 milliseconds before neuron B. The job of a synaptic learning rule is to strengthen or weaken synapses, according to the pattern of activity of the presynaptic and postsynaptic neurons. In this case STDP will preferentially strengthen the $A{\to}B$ synapse and weaken the $B{\to}A$ synapse. It took neuroscientists a surprisingly long time to hit upon this simple learning rule. It was only in the 1990s that STDP was conclusively demonstrated.[14] Hume would have approved the rule implements a neural cause-and-effect detector. If neuron A fires before neuron B fires, it likely contributed to the firing of B—so this synapse is strengthened. Whereas the $B{\to}A$ synapse is always wasting its breath—like someone always reminding you to lock the door after you've already locked the door—so it is weakened (and may eventually disappear all together).

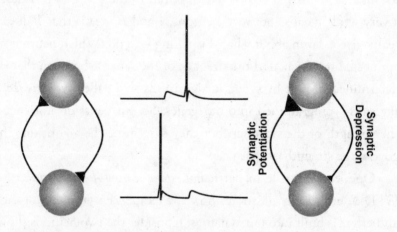

Figure 2.2: Spike-timing-dependent plasticity. Two neurons reciprocally connected to each other by two synapses (represented by the black triangles). If the lower neuron consistently fires before the upper one, the synapse from the lower to the upper neuron will get stronger (synaptic potentiation), and the synapse from the upper to lower neuron will get weaker (synaptic depression).

It is believed that the ability of synapses to learn cause-and-effect relationships between neurons is in part responsible for the brain's ability to learn relationships between events in the external world. In our example, the STDP learning rule may help create neurons respond to the sequence *z-o-e*, but not the rarely heard *e-o-z*—and thus help Zoe learn to recognize her name. But STDP is simply one of many learning rules in the brain's arsenal. Indeed, STDP operates near the finest temporal resolution of the nervous system—a difference of a few milliseconds in the timing of a postsynaptic spike can determine whether a synapse becomes weaker or stronger. STDP cannot capture the relationship between events separated by seconds and beyond. For this, more complex mechanisms based not on two isolated neurons but on multiple populations of neurons are needed. One way or the other, however, the neurons and synapses within our brains manage to connect the dots between events separated by short and long intervals, allowing us to make sense of the events unfolding around us.

TELLING TIME ACROSS SCALES

Close your eyes and focus your attention on some sound in your environment—perhaps the hum of an appliance. You can easily tell if the sound is coming from your left or your right. But how does your brain figure out where in space the sound is originating? A sound coming from the left takes a bit more time to arrive at your right ear than your left. These so-called interaural time delays are a function of the speed of sound and the size of your head. For humans, detectable delays can be around ten microseconds—one thousand times less than the resolution of the chronometers used to time the hundred-meter sprint in the Olympics. The parts of your brain responsible for processing sound must measure these delays to calculate the location of sound sources. Evolution has exploited the fact that because the speed of sound is fairly constant, space and time are complementary—thus telling time allows us to "tell" space.

It is on a slightly longer time scale—from tens of milliseconds to around a second—that our ability to tell time is at its most impressive. Within this range, we can not only estimate the interval between two events in time, but also parse and interpret the complex temporal patterns of music and speech. For example, the duration of syllables or pauses in speech help mark the boundaries between words, such as *grade A* versus *gray day*. The duration of words and the speed of speech also contribute to *prosody*, which conveys the emotional state of a speaker—consider the sluggish speech pattern of a clinically depressed individual versus the brisk delivery of an excited teenager. The same is true in music. As the terms *grave* and *allegro* imply, slow and fast musical tempos can be used to convey sadness and happiness, respectively. Much like our ability to see a face in the relationship between the dots of a Seurat painting, we are able to grasp the whole from the temporal relationship between the parts of speech or music. But we can only detect such temporal patterns on the very narrow time scale of around a second. If you slow speech down too much, it becomes unintelligible, and if you speed a musical piece up too much, it ceases to be music (chapter 5).

Telling time is distinct from the process of consciously perceiving the passage of time. Consciousness is simply too slow to provide a real-time account of the pauses between words, or to count down to the moment we should reach out and catch a ball. But on the scale of seconds and longer, we are aware not only of the flow of time but have a rough sense of how much time has elapsed between different events. We can consciously anticipate when the red light will turn green. We sense that a stream of TV commercials is about to end and that the game is about to recommence. And we figuratively count the seconds as we eagerly wait for the gentleman in line ahead of us to decide if he wants fries with that.

The brain is a product of natural selection, and was thus "designed" to survive in a harsh and continuously changing world. As it turns out, one of the best ways to prosper in such a world is to be able to predict *what* will happen in the future, and *when* it will happen. So the brain is both an anticipation machine and a machine that tells time. It quantifies the passage of time across a range of over twelve orders of magnitude—from the tiny difference in the time it takes a sound to arrive at the right ear versus the left ear, to the ability of some animals to anticipate the seasons.

We are surrounded by the clocks on our wrists, smartphones, cars, appliances, walls, and computers. But it turns out that we are not only surrounded by clocks, we are also filled with them. The brains and bodies of humans and other animals measure time—even an individual liver cell can tell the time of day. But how does the brain tell time? What part of the brain tells time? We now know there is no single answer to these questions. Evolution has endowed the brain with a multitude of mechanisms to tell time. This different-clocks-for-different-time-scales strategy—which I will refer to as the *multiple clock principle*—stands in contrast to man-made clocks. Even the simplest of digital wristwatches can accurately measure hundredths of a second, seconds, minutes, hours, days, and months. In the brain, however, the neural circuits responsible for timing Beethoven's *Fifth* have no hour hand, and the circuits that govern our sleep-wake cycle have no second hand. While this is perhaps counterintuitive at first, we will see that, given the fundamental importance of time to every aspect of behavior and cognition and the distinct set of temporal problems the brain must solve, it is exactly what we should expect.

3:00 DAY AND NIGHT

Maybe it is just as well if we face the fact that time is one of
the things we probably cannot define. . . . What really matters
anyways is not how we define time, but how we measure it.

—RICHARD FEYNMAN

One of the most insignificant mysteries in all of science is: *why do
mice love running wheels*? Anyone who has had a pet mouse or rat,
or observed one in a pet shop, has likely observed their unbounded
enthusiasm for running in a wheel. But why do they run? It does not
seem to be simply because the poor guys have nothing better to do.
People tell stories of finding wild mice spinning away on running
wheels abandoned in a garage and of lab rats that managed to escape
from their cages, only to get caught again because they decided to
take a spin on a running wheel. These anecdotal observations have
been backed by a study in which biologists placed running wheels and
hidden cameras in natural habitats and observed wild mice running
on the wheel, jumping off, and jumping back on again.[1] Like teenag-
ers spending their hard-earned money in an arcade, rodents are even
willing to "work" to run. When running wheels are rigged with a
brake, rats will press a lever to release the brake in order to take a
spin.[2] There is also a dark side to wheel running. When rats are kept

on restricted diets, running wheels can be detrimental to their health. Rats with limited access to food will increase their wheel running and, compared to rats given the same amount of food but no access to a running wheel, exhibit more health problems and an increased mortality rate.[3]

Whether or not we ever resolve this minuscule mystery, the fact that mice, rats, and hamsters are compulsive wheel runners has greatly advanced our understanding of how the brain tells time—or at least the time of day. The graph displayed in Figure 3.1 is called an acto-gram. It captures the pattern of wheel running of a mouse by plotting a vertical tick mark each time the mouse-propelled wheel completes a full revolution. To provide better visualization and avoid a break in the continuity of the plot during any 24-hour period, the graph is *double-plotted*, meaning that the activity on consecutive days is plotted both to the right of, and below, the previous day. The black-and-white bar on the top of the plot represents the 24-hour lights-on/lights-off cycle in the room. Mice and rats are nocturnal creatures, thus they prefer to run during the nighttime—although in the lab their "night-time" might be our daytime because chronobiologists often flip the light-dark cycle of the rooms where mice are kept so that graduate students don't have to stay up all night long to study them. The plot shows that when the lights go out, the mouse jumps on the wheel and starts running, jumping on and off the running wheel the entire night. After a few days, the investigators switched to permanent dark-ness. We can see that even in the absence of any external cues about whether it is "day" or "night," the mouse continues to exhibit a robust rhythm, oscillating between activity and rest. But in constant dark-ness something interesting starts to happen: the period of this cycle drifts from the normal 24-hour day to one with a shorter period—as indicated by the progressive shift to the left. Thus, the sleep-wake rhythm, of mice at least, is not tuned to a precise 24-hour cycle.

For millennia it was thought that the daily fluctuations in sleep and activity of humans and other animals was governed by external

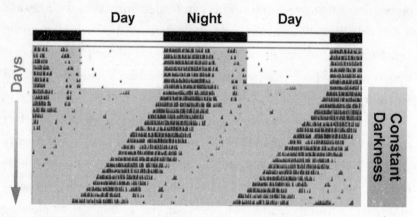

Figure 3.1: Running wheels and actograms. The nocturnal activity of a mouse is indicated by the black tick marks, which represent the revolutions of a running wheel. If mice are kept in constant darkness, their circadian rhythm continues with a period of approximately 23.5 hours, resulting in a progressive leftward shift of the activity pattern. Actograms are double- plotted, meaning that the same 24-hour period is represented at the end of a row and the beginning of the row below it. (Modified from Yang et al., 2012 under CC BY license)

cues, most importantly sunrise and sunset. But experiments similar to those shown in Figure 3.1 established that even in the absence of external cues, animals continue to exhibit daily oscillations in their sleep, activity, eating, and body temperature. These cycles prove that there must be some internal clock—a *circadian clock* (*circa* meaning approximately, and *dian* meaning day)—governing the daily rhythms of our lives.

How good is the circadian clock, and how does it compare to man-made clocks? The performance of clocks, whether of the biological or man-made variety, can be measured by both their precision and their accuracy. Precision refers to the average deviation over many cycles of the oscillator, while accuracy refers to how close the average period is to some target or desired period. If the swing of a pendulum should be 1 sec, but its mean period is 0.8 seconds, it is not very accurate (off by 20 percent). But if over tens of thousands of swings the minimal and maximal period remain between 0.79999 and 0.80001 seconds, it is nevertheless very precise. As can be seen from Figure 3.1, the period of the circadian clock is not exactly 24 hours, but naturally cycles with a period closer to 23.5 hours.[4] Thus, in relation to the time it takes our planet to complete a single spin, the circadian clock is reasonably accurate—a period of 23.5 hours is off by 2 percent. Nocturnal animals generally have circadian clocks with a period shorter than 24 hours, while diurnal creatures, such as humans, tend to have circadian clocks with intrinsic periods slightly longer than 24 hours. The precision of the circadian clock is more impressive. We can see this in Figure 3.1 by noting that the shift in activity to earlier and earlier times is more or less the same each day (different rows). Studies show that across days in constant darkness, the standard deviation of when the mouse starts running can be as low as 10 to 20 minutes, a precision of approximately 1 percent of the clock's 23.5 hour period.[5]

It is this impressive precision of the circadian clock that presumably contributes to the ability of some people to awaken at approximately the desired time. William James referred to the ability to self-awaken in his magnum opus, *The Principles of Psychology*: "All my life I have been struck by the accuracy with which I will wake at the same exact minute night after night and morning after morning." Under laboratory conditions, however, people's ability to self-awaken is rarely as accurate as they believe it to be—indeed it is likely that self-awakening relies in part on the sleeping brain's ability pick up some external cues.[6] Nevertheless, as we will see in chapter 7, a precision of 1 percent surpasses that of all man-made clocks up to the seventeenth century, when Christiaan Huygens figured out how to create the first high-precision pendulum clocks.

ISOLATION EXPERIMENTS

Circadian rhythms observed in the absence of any external signals are said to be *free-running* rhythms. Studying free-running circadian rhythms in humans, however, requires finding people who are willing to submit themselves to total isolation from the external world for many days or even months at a time. In one of the most famous of such experiments the French geologist Michel Siffre spent six months in a cave in Texas in 1972. The experiment was backed by NASA, which foresaw a need to understand the effects of chronic isolation on the body and mind for potential interplanetary missions. Deep inside the cave, Siffre was supplied with plenty of food and water, a simple base camp, and equipment that recorded his sleep patterns. There were no externally driven changes in light, or significant temperature fluctuations to give him hints of the time. He was "free running," but unlike the free-running lab mice, he was not in constant darkness; he could phone the surface unit at any time to have a bank of lights in the cave turned on or off.

Isolation experiments in mice do not seem to produce abnormally high levels of physiological stress or significant mental distress. It is not hard to imagine a wild mouse, living in a cave or a basement, isolated from any sources of light for days at a time. Additionally, vision is not as important for rodents as it is for humans—as nocturnal animals, mice and rats rely heavily on hearing and a highly sophisticated whisker array to navigate the world. Isolation experiments are much more taxing on humans. Indeed, Siffre suffered through bouts of depression, mental lapses, forgetfulness, and thoughts of suicide, and his circadian clock went batty. Over the first few days his circadian cycle extended to 25 or 26 hours, but over time it jumped around a lot, at times shifting to a 48-hour period in which Siffre slept for 16 hours and stayed awake for over 32 hours.

On day 179, Siffre was notified that the isolation experiment had ended. This caught him by surprise, because, by his count, he had been in the cave for 151 days—off by 16 percent. In essence, time had dilated, as his personal sense of time had slowed down compared to objective time. Given the soul-crushing dullness of his isolation, it is hard to imagine that he underestimated the amount of elapsed time. But this form of time dilation has been observed in a number of human isolation experiments. In 1988 Véronique Le Guen spent 111 days in isolation in a cave in France, and at the time of her exit she thought 42 days had elapsed! In 1989 an Italian interior decorator, Stefania Follini, spent four months in a cave fifty feet below the surface of the Earth. Almost four months in, she believed only two months had elapsed. In 1993 an Italian sociologist spent a year in isolation in a cave, and when he exited on December 5, he believed it to be June 6.[7]

A limitation of these experiments is that subjects may not be as isolated from circadian clues as it may seem. Caves have their own biome, including bats and insects that may provide a cave dweller some conscious or unconscious clues about external time. Siffre, for example, recounts his failed attempts at seeking some companionship with a cave mouse, and presumably the mouse was more likely to show up at night. To address these limitations, and to void the need for subjects and scientists alike to live for extended periods in remote areas, chronobiologists also performed isolation experiments in specialized labs or bunkers. A study published in 1985 examined forty-two volunteers who had been in isolation for periods between a week and a month. Subjects lived alone in a bunker and received no information about the actual time from the outside world. They prepared their own food and could turn the lights on and off at will. They had to report their sleep and wake times, and their body temperature was continuously monitored. Again, the majority of the participants believed that the experiment lasted 20–40 percent less time than it actually did.[8] And like the cave experiments—and in contrast to the rodent studies—the circadian period of the subjects often jumped around, rather than settling into precise and reproducible cycles.

Sleep-wake cycles are not the only way to measure what the circadian clock is up to. Many physiological measures fluctuate according to the time of day. Human body temperature, for example, is not actually a constant 98.6°F. It fluctuates around this average over the course of a day, generally peaking in the early evening. In many subjects the temperature rhythm remained close to 24 hours, even when their sleep-wake cycle was as short as 20 hours or as long as 40. Providing an important clue that we have more than one circadian clock within us and that these clocks may not always agree with each other.

THE SUPRACHIASMATIC NUCLEUS

Sitting at the bottom of your brain is a structure called the hypothalamus. And at the bottom of the hypothalamus, hovering above the intersection of the nerves carrying information from your left and right eyes—the so-called *optic chiasm*—lies the appropriately named *suprachiasmatic nucleus*.

Since the 1970s we've known that rodents with a lesioned suprachiasmatic nucleus exhibit sleep patterns that are anything but circadian. They sleep in short bouts peppered throughout the day and night. These early observations led to the hypothesis that the suprachiasmatic nucleus was the master circadian clock. Proof arrived in the 1980s via a series of converging experiments. One of the most compelling was, in essence, a brain transplant experiment.[9] Free-running hamsters have a sleep-wake period very close to 24 hours. There is a mutation, however, that results in hamsters with a much shorter free-running "day" of 20 hours. If the suprachiasmatic nucleus is the master circadian clock, researchers reasoned, it would be possible to transform a 20-hour hamster into a 24-hour hamster by transplanting the suprachiasmatic nucleus from one strain of hamster to the other. Generally speaking, such brain-area transplants would be limited to science fiction. But the relative simplicity of the suprachiasmatic nucleus makes it one of the few parts of the brain that one could effectively transplant. Unlike many brain areas, the suprachiasmatic nucleus is a fairly compartmentalized structure—it receives inputs from few other areas of the brain. And, more importantly, it communicates with the rest of the brain not only by electrical impulses mediated through delicate axons, which do not regenerate very well, but by releasing hormones directly into the bloodstream. When the investigators lesioned the suprachiasmatic nucleus of a host and transplanted cells from a hamster of one strain to the other, they transformed the short-day hamsters into long-day hamsters, and vice versa. The circadian rhythm of the suprachiasmatic nucleus was not governed by the host's body or brain; rather, it was the suprachiasmatic neurons—a humble collection of 10,000 neurons or so—that took control and told the host's brain when to go to bed and when to get up and start wheel running.

CELLS THAT TELL TIME

Is having a brain a prerequisite for having a circadian rhythm? Tracking and anticipating the light and temperature fluctuations imposed by the rotation of the Earth is so important that virtually all forms of life have circadian clocks. Indeed, the first free-running circadian clock experiment was performed in the plant *Mimosa pudica* ("touch-me-not"), which opens its leaves during the day to expose them to sun and closes them at night. In 1729 the French astronomer Jean-Jacques d'Ortous de Mairan placed a *Mimosa* in a pitch-black room and noted that the leaves continued to open and close in synch with external time over the course of many days. Mairan himself didn't seem to believe his own results. During Mairan's time one of the most pressing scientific challenges was to tell time at sea, so it was difficult for scientists of that era to accept that a lowly plant came with a built-in clock. Mairan assumed the "behavior" of the *Mimosa* must be guided by some other signal, such as temperature, or some unknown magnetic field that was telling the plant when to open and close its leaves. It took more than two centuries for scientists to understand that all plants and animals have their own personal clocks, and that even individual cells could oscillate with a 24-hour period.

When we say a single cell oscillates, we do not mean it is physically vibrating like a quartz crystal, much less swinging back and forth like a pendulum. Rather, the oscillations refer to the concentrations of proteins within a cell. Cells are not static entities. Depending on what they are currently working on, the concentrations of different proteins within a cell change dramatically. For instance, the cells lining your intestine will ramp up production of digestive enzymes during a meal. Similarly, cells in your pancreas will increase the synthesis of the proteins necessary for the production of insulin when there is an increase in glucose in the bloodstream. Cells are also not simply switches governed by external stimuli; they have their own internal rhythms. Much like mice, single cells can also free run. When kept at a constant temperature in an unchanging biochemical milieu, many cells have their own private circadian rhythm, as measured by the fact that the concentration of some proteins rises and falls with a period of approximately 24 hours. These cellular oscillations can be strikingly visualized by a bit of clever genetic engineering. Fireflies emit light because they make the enzyme luciferase, which in the presence of the appropriate substrate (a small molecule called luciferin) releases energy in the form of photons. Scientists have inserted the luciferase gene in cells ranging from bacteria to mold, plants, fibroblasts, and, of course, suprachiasmatic nucleus neurons. When the transcription of the luciferase gene is placed under the control of a protein that naturally exhibits a circadian rhythm, the concentration of luciferase within the cell will also oscillate. The result is that cells literally light up and then slowly fade away, only to slowly light up again approximately 24 hours later.

How does a single bacterial cell keep track of the time of day? Before answering this question, it is worth pointing out, that an equally valid question might be: *why would bacteria care what time it is?*

THE FIRST CLOCK

We will see in chapter 7 that without the widespread availability of accurate man-made clocks, the industrial revolution would not have been possible. Assembly lines, in which specialized laborers performed serial steps in a manufacturing process, required the temporal coordination of large numbers of workers. But a bit before the industrial revolution—a billion years or so before—evolution had already created factories and solved the problem of coordinating different processes over time. The most important assembly line on planet Earth is photosynthesis: the series of biochemical steps that usher the energy of solar photons through a line of proteins in order to create stable, energy-rich biomolecules, the most famous of which is glucose.

Cyanobacteria are photosynthetic organisms, and photosynthesis is the ultimate daytime job. Just as a factory owner wouldn't pay employees to sit in a factory overnight doing nothing, a cyanobacterium would be wasting energy synthesizing the proteins needed for photosynthesis during the night. It would, however, be advantageous to ensure that these molecules are ready to get to work before first light in order to fully exploit the sun's energy. Evolution's solution, of course, was an internal alarm clock that anticipates sunrise. Thus, one of the driving forces for the evolution of circadian clocks was the highly adaptive coordination of cellular functions with cycles of light and dark produced by the Earth's rotation.

The evolutionary advantage of telling the time of day—that is, of having a good circadian clock—has been beautifully demonstrated in an elegant experiment that pitted different strains of cyanobacteria against each other. Figure 3.2 illustrates the circadian rhythm of the two strains used in the experiment, one with a short period of around 23 hours, and one with a long period of approximately 30 hours. Investigators placed both strains in the same petri dish and asked if one strain would eventually take over the whole petri dish. The clever part of the experiment was to do this under two conditions: one in which the lights were turned on and off every 11 hours for an artificial day of 22 hours (close to the natural period of the 23-hour cyanobacteria), and in the other every 15 hours to simulate a 30-hour day. The researchers found that after a month the cultures kept on a 22-hour cycle were dominated by the short-period strain. In contrast, when the cultures were kept on a 30-hour cycle, the long period strain was the victor.[10] A rhythm of 22 hours in a 30-hour day, or of 30 hours in a 22-hour day, resulted in cellular rhythms that were always going in and out of phase with the light and that were less efficient at extracting energy from light. Therefore, it is not sufficient to have a circadian clock—the period of the clock must resonate with the natural cycle of the environment in order to provide an evolutionary advantage.

Optimizing photosynthesis is one reason single-cell organisms benefit from having a circadian clock, but it may not have been the

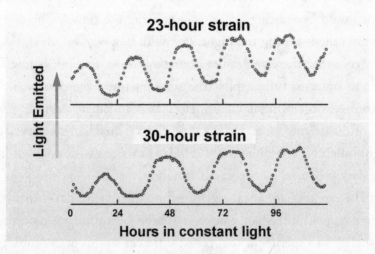

Figure 3.2: Fast and slow circadian rhythms in cyanobacteria. The circadian rhythms of two strains of cyanobacteria with periods of approximately 23 and 30 hours. The bacteria were genetically engineered to emit light in a manner proportional to the concentration of a specific protein. When these strains are forced to compete with each other for resources in an environment with a 23-hour light-dark, the 23-hour strain will win; in contrast, if they are placed in a 30-hour light-dark cycle, the 30-hour strain will win. (Adapted with permission from Johnson et al., 1998)

first. Just as fundamental to life is the ability to divide and reproduce. And a key event of cell division is the replication of DNA—the codex in which the recipe for life itself is written down. DNA replication is notoriously sensitive to ultraviolet (UV) radiation, which is why repeated sunburns are a risk factor for skin cancer in humans and why the label on your sunscreen touts its UV absorption properties. The dangers of UV radiation are much more severe for single-cell organisms that do not have the benefit of a protective organ such as the skin, full of the UV-absorbing pigment melanin. A cell that divides under UV light risks damaging its DNA, a danger that recedes at night. Some chronobiologists thus endorse the so-called *escape from light* hypothesis, which posits that the original driving force for the evolution of circadian clocks was to help cells divide at night.[11]

THE MECHANICS OF THE CIRCADIAN CLOCK

Now back to the main question: *how does a single cell perform the feat of accurately keeping track of the time of day?* The initial step toward answering this question was taken at the California Institute of Technology by the Nobel laureate Seymour Benzer and his student Ron Konopka in the early 1970s.[12] Benzer's lab was studying the famous fruit fly *Drosophila melanogaster,* which, like all flies, start their lives as larvae encased in a pupa, from which they emerge a few days later in their adult form. The process of breaking free from the pupa is well timed. They emerge, or *eclose*, during the dewy early morning hours to avoid sun-induced dehydration. Konopka wanted to understand the genetics of the circadian clock by finding mutations that caused larvae to eclose at the wrong times. He identified three mutants: one that emerged at more or less random times, one that emerged early, and one that emerged late. When placed in constant darkness the activity patterns of these mutants was similarly affected: the adults were either active at random intervals throughout the day, exhibited a free-running cycle of only 19 hours, or exhibited an abnormally long cycle of 28 hours. Konopka was confident all three mutations were on the same gene, a gene he called *Period*. Over a decade later, a research group led by Michael Rosbash identified and sequenced the *Period* gene.[13] Subsequent work identified numerous additional genes critical to the circadian clock—genes that often go by temporally evocative names such as *Clock*, *Cycle*, and *Timeless*.

The details of the how these genes and their protein products interact to create a robust and highly reliable circadian clock are quite complex, but the general underlying principle is simple. So simple, in fact, that one can see this principle in operation by looking at a toilet tank. Every time a toilet is flushed a negative feedback loop is engaged to refill the tank while not allowing it to overflow. The low level of water causes the *floater* (a "ball" that is connected to a valve) to go down, opening the water valve in the process, and as the water rises, the floater moves back up and closes the valve. If one were to purposely put a small hole in the tank, allowing the water to slowly leak out, the floater would eventually fall enough to reopen the water influx valve and refill the tank. The result would be an oscillation: the slow drop in water level eventually opens the water valve, producing a rise in the water level, which turns the water influx off. The water then proceeds to slowly leak out again, and the process repeats. Indeed, if your toilet tank mysteriously starts making noises in the middle of the night, it is probably already oscillating as result of a leak in the flush valve.

The circadian clock is, of course, way more complicated than your toilet, but the idea is the same, even though the mechanism governing the period of the circadian clock is a mouthful: *transcription/translation autoregulatory feedback loop*: *Transcription* because genes encoded in DNA are transcribed into RNA. *Translation* because these strands of RNA are translated into proteins. And *autoregulatory feedback loop* because these proteins inhibit further transcription of the very genes that led to their synthesis (thus closing the water valve). One of these proteins is called Period, the product of the *Period* gene. As the concentration of Period increases, it eventually turns off the gene that synthesizes it. Then, as it is slowly degraded, the *Period* gene kicks back into action, and the concentration of Period will increase again. Guess how long this cycle takes?

A circadian clock requires much more than simply oscillating at a frequency of approximately 24 hours—its oscillations have to be robust. Like the clockmakers of the eighteenth century who struggled with the effects of temperature on pendulum and mechanical clocks, evolution had to overcome the problem that the speed of biochemical reactions changes with temperature. We still do not fully understand how ectothermic organisms, such as cyanobacteria, plants, and flies, maintain a period of approximately 24 hours over daily and seasonal temperature fluctuations. But we know that there are a lot of additional proteins and genes that interact with the basic molecular machinery of the transcription/translation autoregulatory feedback loop, and some of these bells and whistles likely contribute to temperature compensation.[14]

JET LAG

Radio station WWVB, located near Fort Collins, Colorado, is an important, albeit mind-numbing, one. All day long it sends out the *Coordinated Universal Time* to clocks and watches across North America—specifically, to those clocks and watches that can be synched through radio signals. The suprachiasmatic nucleus has a similar job. The molecular machinery that makes up the circadian clock is present in most mammalian cells. Thus, the concentration of the Period protein is oscillating away not only in your suprachiasmatic neurons, but throughout most of the cells in your body.[15] It is the job of the suprachiasmatic nucleus to keep them all in synch.

During the day suprachiasmatic neurons ramp up their activity, sending signals to downstream areas of the nervous system.[16] The level of neural activity in these neurons indirectly translates into whether we are wide awake or sleepy, and serves as a calibration signal that provides information about the time of day—or at least what time the suprachiasmatic nucleus thinks it is. Like most cells in your body, suprachiasmatic neurons live in perpetual darkness, deep within the recesses of the cranium. Which raises the question: *how does the suprachiasmatic know the correct external time—that is, if it is day or night?*

In chronobiological lingo, the suprachiasmatic nucleus has to be *entrained* by external cues, the *zeitgebers* (time givers). Sunlight is, of course, the most important zeitgeber. The location of the suprachiasmatic nucleus at the intersection of the left and right optic nerves is not coincidental; this location makes it ideally situated to receive raw data about whether it is light or dark outside the skull. This information helps entrain the circadian clock, and ensures that the body's internal rhythms are in proper phase with Earth's rotation. But entrainment is easier said than done.

Anyone who has endured the mental and physical haziness of transmeridian travel can attest to the challenges of resetting our circadian clocks. It can take days to resynch the suprachiasmatic nucleus after a trip between Los Angeles and London—the rule of thumb is up to a day for each hour your circadian clock has to be advanced. In contrast, the clocks on our wrists can be instantly reset to match the local time zone. This difference reflects the distinct design principles of man-made versus circadian clocks. The oscillation frequency of the quartz crystal of most wristwatches is 32,768 Hz, so an hour simply consists of counting 32,768 x 60 x 60 of these ticks. Resetting your watch upon arriving in London does not require tinkering with the oscillator itself, but simply changing the current total tick count by advancing the hour hand (or digit)—a mere formality. In sharp contrast, the pendulum of the circadian clock requires a full day for a single swing, so resetting it requires the much more delicate task of mucking with the actual oscillator, akin to advancing a pendulum in mid-swing. The "swing," in this case, refers to the rise and fall of the circadian proteins within our cells. It is simply not possible to instantly reset the concentration of the circadian proteins within a cell, any more than it is possible to instantly reset an hourglass that is half full.

At the equator, a 1-hour time zone (15 degrees of longitude) corresponds to a distance of approximately 1,600 km. So to cross the eight time zones equivalent to a Los Angeles-to-London trip in a 12-hour period, it is necessary to travel at an average speed of above 1,000 km/hr—far faster than the speeds any animal can run, swim, or fly. Thus jet lag is a uniquely modern condition, one that results not only in cranky tourists and groggy academics at international scientific conferences, but in poor decision making by airline pilots, military personnel, and diplomats. All jet lag, however, is not created equal. Eastward travel is significantly harder to adjust to than westward travel. Eastward travel requires a phase advance of our circadian clock—when traveling from Los Angeles to New York we must set our wristwatch forward three hours—whereas westward travel requires a phase delay. Traveling from the west to the east coast is akin to going to bed early, whereas an east-to-west trip is akin to staying up late—and most people find it harder to go to bed early than to stay up late. Consistent with this intuitive view, jet lag is generally more severe with eastward travel because circadian clocks are harder to phase advance than delay, although the mechanistic reasons for this are not fully understood.

Eastward jet lag seems to be harder on mice as well. When old mice are placed in conditions that simulate chronic eastward jet lag, by shifting their light-dark cycle six hours forward every week, death rates are significantly higher after eight weeks compared to mice that have had their light-dark cycle shifted back by six hours (simulating westward travel).[17]

FIGHTING THE CLOCK

Are you a "morning lark," early to bed and early to rise, or a "night owl," late to bed and late to rise? These avian terms refer to different *chronotypes*, and there is a standard diagnostic questionnaire to determine if you are a lark or an owl: it includes questions about when people prefer to go to bed, when they feel most alert, and when they are more likely to exercise. Different chronotypes reflect a natural variation among individuals, which is influenced by environment and age. But even though we may be predisposed to being a lark or an owl, most of us can adapt—albeit grumpily—to a range of different workday schedules. There are, however, people who have so much trouble not dozing off at 8:00 p.m. that it prevents them from carrying out normal social and professional activities. Such individuals are said to have a circadian rhythm sleep disorder. In the late 1990s it was discovered that there was a genetic basis to some of these disorders through the study of five generations of a family with a number of extreme morning larks. At least one member of the family—the one willing to submit to an eighteen-day free-running isolation experiment—exhibited a sleep-wake period of close to 23 hours, in contrast to the standard period of a bit over 24 hours in humans. In 2001 researchers pinpointed the genetic mutation associated with *familial advanced sleep-phase disorder*. In a striking validation of decades of fundamental research in flies and rodents, it turned out that the first identified gene associated with a human circadian disorder was the same gene that Benzer and Konopka identified in the 1970s as critical to the circadian rhythms of fruit flies: the *Period* gene.[18]

People with familial advanced sleep-phase syndrome are living with a 23-hour clock in a 24-hour world. They are continually fighting against their own internal clock. But one does not need to have a mutation on the *Period* gene to be engaged in a similar clock struggle. In our modern 24/7 world, a significant percentage of the workforce are shift workers: factory workers, pilots, nurses, doctors, and police officers who work at night and do their best to sleep during the day. The sleep-wake cycle of shift workers is generally out of synch with their natural rhythm. Compounding the problem, most shift workers continuously adjust their cycle: they are nocturnal beings during their workweek and diurnal on the weekend. Not surprisingly, shift work is a risk factor for a number of health problems, including ulcers, cardio-vascular disease, and type 2 diabetes. The underlying causes for these problems are not fully understood, but they are in part a product of the mismatch between the internal physiological cycles and external stimuli. For example, hormones such as insulin generally increase in anticipation of normal meal times. A chronic mismatch between when the body expects to be fed and when it is actually fed seems to contribute to diabetes.[19] Animal studies, for example, have shown that genetically deleting the circadian clock of the pancreas—while leaving the clocks in the suprachiasmatic nucleus and other organs intact—increases the incidence of diabetes in mice. This suggests that the coordination between the many circadian clocks within our body is critical to healthy physiology.[20]

The detrimental effect of living outside one's natural rhythm has been well demonstrated and raises the question: *might it be better to have no clock at all than a clock that is perpetually out of synch?* Surprisingly, the answer may be *yes*. As mentioned above, hamsters with circadian clock mutations can have free-running periods significantly below or above 24 hours. One such mutation results in a strain with an intrinsic circadian period of 22 hours. Compared to their wild-type counterparts these mutants have a shorter lifespan when living in a 24-hour world. Lesioning the suprachiasmatic nucleus of these animals actually prolonged their lives. This is a fascinating example of a situation in which part of the brain seems to be doing more harm than good.[21]

The ability to accurately schedule physiological functions and anticipate the sunrise and daily meals is a valuable adaptation, but if the circadian clock does not resonate with the period of the world around us, the effects are so serious we might be better off without a clock altogether. In the distant future humans may colonize other planets, and it is highly unlikely that the rotation period of any Goldilocks planet will resonate with our own circadian clock. Mars, for example, has a close-enough rotation of 24 hours and 39 minutes, but a "day" on Mercury lasts over fifty-eight Earth days. Thus, if and when the time comes, we may find that best way not to continuously fight our own clock will be to turn it off altogether.

THE MULTIPLE CLOCK PRINCIPLE

The atomic clocks at the National Institute of Standards and Technology are astounding devices not only because of their unfathomable precision, but because they are used to track time across temporal scales: from nanoseconds to years. Does the circadian clock within the suprachiasmatic nucleus tell time across different scales? Is it responsible for our ability to distinguish a half musical note from a full note, figure out if our order at the coffee shop is taking an unacceptably long time, or govern the twenty-eight-day reproductive cycle of women?

One of the early approaches toward answering this question was to ask if changes in the duration of the circadian period produced during free-running isolation experiments altered people's ability to time shorter intervals. In one set of experiments, the volunteers were asked to press a button every time they thought one hour had elapsed. During a 16-hour awake period, one subject estimated an hour to be approximately 2 hours, and during a 44-hour awake period, his estimate of one hour was close to 3.5 hours. Overall, there was a correlation between the length of individuals' circadian period and their estimate of an hour. From this one might conclude that the circadian clock is used to tell all intervals of time, including shorter intervals like the beat of a song or the duration of a traffic light. This, however, is not the case. When asked to press a button for durations ranging from 10 to 120 seconds, there was no significant relationship with the duration of the individual's sleep-wake cycle. The absence of any relationship between the circadian period and temporal judgments on the scale of seconds to a few minutes is consistent with a large body of work that shows we have distinct circuits devoted to telling time across different scales.[22] For example, experiments in rodents reveal that the mutations that perturb the circadian clock, or lesions to the suprachiasmatic nucleus, do not alter their ability to time events on the scale of seconds.[23] (In the next chapter we will see how exactly scientists ask animals how much time they think has passed).

Our understanding of how circadian clocks tell time also assures us that they do not contribute to our ability to tell time on the scale of seconds. The building blocks and principles of operation of a circadian clock render it incapable of timing short intervals. In other words, the circadian clock does not have a minute hand, much less a second hand. The biochemistry of translation/transcription feedback loops is simply too slow to be of any use when attempting to determine if the red light is about to change. This is not to say that the circadian clock cannot influence timing on other scales: it can. But only because circadian rhythms indirectly affect pretty much all physiological and cognitive functions, including learning, memory, reaction time, and attention—which is why it is best not to have jet-lagged pilots flying jets, or sleep-deprived truck drivers on our freeways.[24]

What about longer intervals? Do circadian clocks contribute to timing of slower rhythms—so-called *infradian* rhythms? The moon revolves around the Earth with a cycle of approximately 29.5 days. Historically, this cycle has left a profound imprint on human culture. Most calendar systems, including our current Gregorian calendar, are based on the approximately twelve full moons that occur throughout a year. Etymologically the word *month* traces back to *moon*. It has long been hypothesized that the phases of the moon play a role in human physiology. For example, as the term *lunacy* suggests, it was believed that the full moon could cause people to go insane; contemporary scholars think this association arose because changes in sleep patterns triggered by the light of the full moon may have pushed people suffering from epilepsy or bipolar disorder over the edge. Additionally, the fact that the human menstrual cycle is very close to the lunar month hints at a role for the moon in human reproduction. This appears to be a mere coincidence, as the menstrual cycle of other primates can be significantly shorter or longer. So, other than the obvious fact that lunar light could alter sleep and social activities, there is little evidence that the phases of the moon have any direct impact on human physiology.[25]

The moon does, however, play an important role in the physiology of many animals. Most notably, some marine invertebrates synchronize their development and mating behavior to lunar cycles. In natural environments the moon is the primary source of light at night, and a full moon enhances the ability of predators to see potential prey, so the most vulnerable phases of the life cycles of some animals occur out of phase with the full moon. Other animals use the phase of the moon to synchronize mating. For species with internal fertilization, sexual reproduction requires that a female and male be in the same place at the same time. This is not necessarily a requirement for species with external fertilization; nevertheless, it is important that males and females spawn at approximately the same time. Sea worms, segmented invertebrates related to earthworms, are one of the species that rely on the phase of the moon during the breeding season as a synching signal to maximize the chances of eggs and sperm hooking up. This synchronized spawning can result in millions of worms coming to the surface at the same time, which in some cultures results in a gastronomical event. Indeed, the natives of some Indonesian islands use the spawning of sea worms to mark the onset of the festivities of the new year.[26]

The sea worm's internal circalunar clock can be demonstrated by the lunar equivalent of free-running circadian experiments. The circalunar clock must first be entrained, not by sunlight, but by moonlight (or in the laboratory by exposure to dim light for a few hours at night). If, after a period of entrainment, worms are kept in the laboratory under a constant day-night cycle, they nevertheless exhibit a 30-day reproductive rhythm. How do these worms keep track of this 30-day cycle? Do they use their circadian clock as a pendulum with a period of 1 day and count up to 30? If this were the case, sabotaging the circadian clock would certainly alter the timing of the reproductive cycle. But this is not the case: when sea worms were given a drug that altered their circadian rhythm, they still maintained a 30-day circalunar cycle.[27] One more piece of evidence in support of our multiple clock principle.

We see that the timing devices within the body and brain are unlike man-made clocks. The same watch can track the milliseconds, seconds, minutes, days, and months of our lives. In contrast, the multiple-clock principle tells us that the brain has different mechanisms to track each of these units of time.

Anticipating the daily changes in light, temperature, and food availability is so fundamental that virtually all life forms, from bacteria to *Homo sapiens,* have high-quality circadian clocks. But the circadian clock is no more suited to time duration of a traffic light than a sundial is to time the 100-meter sprint—it is a one-trick clock.

Time as kept by the circadian clock is not only limited to tracking the hours of the day, but it also hidden from conscious access. Yes, we feel awake or tired depending on the concentrations of certain proteins within the suprachiasmatic nucleus, but we don't *feel* the time of day like we *feel* the heat of midday sun. But we subjectively feel the passage of time, and are keenly aware of the duration of unfolding events. Clearly—as we will see next—the brain has other means to judge the passage of time. Means that transcend the passive measurement of time, and somehow generate a subjective sense of time's passage.

4:00 THE SIXTH SENSE

About 38 years ago, I was on a road in Pennsylvania, I was sleeping in the back of a car. I woke up, the driver of the car was also asleep, the car was veering off the road, the passenger next to her, reached over very slowly, it seemed, grabbed the wheel and pulled that wheel as hard as she could. I can see in my mind what's going on in that car, it was clearly screaming and noise, and I can see the mouths open, but I have no memory of the sound, and she pulls the wheel and the car veers to the right, and very slowly we hit the guard rail, the car flips into the air, and I feel in my gut that all of life is going to change. A memory that must have only taken a second or two, seems like an eternity in my mind. I woke up in the hospital and that's the last time I ever walked.

—JOHN HOCKENBERRY[1]

During life-threatening situations our subjective sense of time can be radically altered, as if shifted into a slow-motion mode. One of the first scholarly reports of this *slow-motion effect* was published by a Swiss geologist, Albert Heim, in 1892. He gathered accounts from members of the Swiss Alpine Club who had experienced serious falls or other near-death events. Ninety-five percent of the group reported what Heim summarized as "a dominant mental quickness and sense of surety. Mental activity became enormous, rising to a hundred-fold velocity or intensity. . . . Time became greatly expanded. The individual acted with lightning-quickness in accord with accurate judgment of his situation. In many cases there followed a sudden review of the individual's entire past."[2]

Review boards for human-subject experiments tend to frown upon putting people in life-threatening situations, so it is difficult to carefully corroborate and study the slow-motion effect. But some studies have asked people to estimate the duration of highly emotional or frightening events, including experiencing an earthquake, watching a scary video, jumping from a height into a net, and skydiving.[3] For the most part these studies confirm that people generally overestimate the duration of the event, which is consistent with reports that external events are unfolding slowly (watching a movie in slow motion takes longer than watching it at normal speed).

In and of itself, however, the overestimation of the duration of emotional events is not particularly surprising because it turns out that there are innumerable perfectly harmless situations in which people also overestimate the passage of time.[4] Indeed, our subjective sense of time is actually quite inaccurate. *A watched pot never boils* and *time flies when you're having fun*, precisely because there are countless circumstances that warp our subjective sense of time. Enduring a very boring lecture or awaiting plane repairs while on the tarmac, for example, can create the feeling of *chronostasis*—the sensation that time is standing still. In contrast, when you are engrossed in a book, immersed in your favorite hobby, or fully engaged in a complex task such as writing computer code, time can seem to vaporize, magically jumping from one moment to another with nothing in between.

What is the relationship between objective clock time and our subjective sense of time? Why does time appear to slow down during life-threatening situations? What is happening in the brain when we say time is flying by, or dragging along? Before we address these questions, we must first distinguish between two distinct types of timing.

PROSPECTIVE AND RETROSPECTIVE TIMING

Telling time is a bidirectional problem. A stopwatch triggered at the start of a marathon provides a continuous measure of how long the marathoners have been running, but it tells us nothing about how much time they spent at the starting lineup waiting for the race to begin, much less about when they got up in the morning. Starting a stopwatch is an example of *prospective timing*: determining the passage of time starting from the present into the future. In contrast, if you walk into a room just in time to see the last grains of sand trickle through the neck of an hourglass, you can deduce something about how much time has elapsed since a past event: an hour ago someone flipped the hourglass over. But unless you flip it over again, the hourglass provides no information about how much time has elapsed since you entered the room. This an example of *retrospective timing*: estimating the passage of time from some moment in the past up until the present.

Throughout the day humans are continuously engaging in prospective and retrospective timing. Consider two scenarios in which you might rely on your ability to estimate temporal durations. First, at a party you are talking to your friends Amy and Bert; Amy asks you to remind her to leave in five minutes because she has somewhere to go. In the second scenario, Amy excuses herself and leaves, and five minutes later Bert asks you, "How long ago did Amy leave?" In both cases you are asked to estimate the amount of elapsed time, but does your brain use the same mechanism to tell time in both cases? No. As far as the brain is concerned, these two timing tasks are fundamentally different from each other. In the first case you know in advance that you will be performing a timing task, you can start a hypothetical stopwatch at $t=0$, and track the passage of time until approximately five minutes have elapsed. But in the second case—where Bert asked how long ago Amy left—your stopwatch is useless because you were never told when to start it. Prospective timing is a true temporal task in that it relies on the brain's timing circuits. In contrast, retrospective timing is in a sense not a timing task at all; it is rather an attempt to infer the passage of time by reconstructing events stored in memory.

The distinction between prospective and retrospective timing explains a few of the mysteries about our subjective sense of time, including what some have called the holiday paradox.[5] A five-hour wait for a delayed plane on your vacation trip to Greece can seem endless as it is unfolding, while an exciting day touring Athens flies by. A week later, however, the airport delay is a mere blip in time, while the busy, fun-filled day in Athens seems quite extended.

This holiday paradox is not an artifact of our modern, fast-paced, high-speed-travel lifestyle. William James wrote in 1890: "In general, a time filled with varied and interesting experiences seems short in passing, but long as we look back. On the other hand, a tract of time empty of experiences seems long in passing, but in retrospect short. A week of travel and sight-seeing may subtend an angle more like three weeks in the memory; and a month of sickness hardly yields more memories than a day."[6]

As they unfold, interesting and engaging activities seem to fly by, in part because we are not thinking about time. So your first tour of the 2,500-year-old Parthenon may fly by, but that five-hour wait in the Atlanta airport will drag along as you continuously check your watch and wonder to yourself *how much longer is this going to take?* Retrospectively, the duration of those activities are estimated in part based on the number of events stored in memory. And since we are much more likely to remember novel and personally meaningful events, the Parthenon is more likely to earn a slot in your memory bank than your first visit to the Atlanta airport bathroom.[7]

The intimate relationship between memory and retrospective timing is strikingly illustrated by the case of the British musicologist Clive Wearing, who developed a severe inability to create new long-term memories after a serious brain infection. While many of his faculties remained intact (including his ability to play music and conduct), he initially spent much of his day writing in his diary "Now I am really completely awake," and later crossing it out, only to write, "Now I am perfectly awake—first time." In the absence of the ability to form new memories, he seemed to be trapped in an infinite loop of an unchanging present. Unable to make sense of where he is or how he got there, the only interpretation his mind can confabulate is that he has perpetually just awoken from sleep. He has no retrospective sense of when he woke up, because he has little or no memory of what happened in the previous minutes and hours.

TIME COMPRESSION AND DILATION

On the scale of seconds, the differences between prospective and retrospective timing can be easily studied, and manipulated in the laboratory. One of the most common ways to surreptitiously alter people's perception of time is by changing the *cognitive load* of the task they are performing. Cognitive load is just a fancy term to describe how easy or hard a task is. In one of the first such experiments, investigators gave subjects a stack of shuffled cards: one group was asked to deal the cards face up into a single pile (low cognitive load), the other group separated the cards into four piles by suit (high cognitive load). All subjects were allowed to deal the cards for 42 seconds. When the subjects knew in advance (the prospective timing condition) they would be asked to verbally estimate the amount of time they were dealing, the average guess was 53 seconds in the single-pile condition, and 31 seconds in the suit condition; in contrast, when subjects did not know they were later going to be asked to estimate the time (the retrospective timing condition), these values were 28 and 33 seconds, respectively. Dozens of subsequent studies have established that prospective timing is strongly modulated by cognitive load: the more complex or challenging a task, the shorter the estimates of how much time was spent performing the task (53 versus 31 seconds). The opposite can happen with retrospective timing: the higher the cognitive load, the longer the task can seem (28 versus 33 seconds). Retrospective timing, however, is not as strongly modulated by cognitive load as prospective timing.[8]

The importance of the difference between prospective and retrospective timing cannot be overstated. For example in the low-cognitive-load condition of the dealing study, subjects were doing the exact same thing (placing cards into a single pile), for the same amount of time; yet the prospective and retrospective estimates were 53 and 28 seconds, respectively. Studies demonstrating large differences in prospective and retrospective time estimates, and the susceptibility of these estimates to cognitive load, also reveal just how inaccurate and unreliable our judgments of elapsed time are. Our subjective sense of time is affected by so many external and internal factors that depending on context the same duration can easily be off by a factor of two. Studies have shown that people tend to overestimate the amount of time they wait in store lines, bank lines, or on hold on the phone by 25 to 100 percent. Indeed, companies subject us to elevator music while we are on hold waiting for customer service because some studies suggest that people report waiting less time if they are listening to music during their wait.[9]

In the laboratory most studies on the distortion of temporal estimates focus on the scale of hundreds of milliseconds to a few seconds. Typically volunteers will sit at a computer and make judgments about the duration of images or sounds. In a study led by the cognitive neuroscientist Virginie van Wassenhove a *reference stimulus* was a static circle displayed on the screen for 500 ms (a half second). The same circle was then presented for either a shorter or longer period of time. Subjects were asked to report whether this *comparison stimulus* was longer or shorter than the reference stimulus by pressing one of two keys. Under such conditions, subjects tend to be fairly accurate—that is, when the comparison stimulus lasts 450 ms they tended to correctly report it as shorter than the 500 ms circle, and when it was 550 ms, they generally reported it as longer than the reference. So the perceived duration of the comparison stimulus is fairly accurate. However, if the comparison stimulus becomes a looming circle—one growing in size—while the reference stimulus remains static, an illusion emerges, a form of chronostasis or time dilation. The looming circle is perceived as lasting longer than a static one; a 450 ms looming circle may be perceived as the same duration as a 500 ms static circle.[10]

An assortment of additional physical features can alter our perception of time on the scale of around a second. For example, auditory stimuli are often perceived as lasting longer than visual ones. Magnitude can also influence temporal judgments: some studies have shown that people will even judge an image of the number 9 to last longer than an image of the number 1 even though they are both displayed for the same duration. Additionally, people perceive novel or unexpected stimuli to last longer than familiar or expected ones.[11]

One of the most common examples of how our sense of time is easily warped is the *stopped clock* illusion. Perhaps you have noticed this illusion upon turning your gaze to an old-fashioned analog clock that has a ticking second hand. Upon shifting your gaze to the clock you may have thought to yourself, "Damn, the clock stopped," but just before you finished that thought, you realized you were mistaken, as the second hand was moving after all. The stopped clock illusion arises because the pause in the movement of the second hand seems to last longer than what your brain thinks a second should last. The illusion seems to be caused by the fact that on the brief scale of a second or less, our own actions, in this case shifting our gaze, can warp our sense of time.[12] It is as if when we shift our attention some internal timer within our brain starts ticking a bit faster, leading to more ticks accumulating within a fixed duration and an overestimation of elapsed time. The stopped clock and other temporal illusions establish that our subjective sense of time is precisely that—not objective.

Figure 4.1: (Paul Noth/The New Yorker Collection/The Cartoon Bank)

CHRONOPHARMACOLOGY

As the cartoon in Figure 4.1 reminds us, our sense of time can be radically influenced by psychoactive drugs. Not surprisingly, this last fact did not escape the attention of William James, who alludes to this through personal experience: "In hashish-intoxication there is a curious increase in the apparent time-perspective. We utter a sentence, and ere [before] the end is reached the beginning seems already to date from indefinitely long ago."[13] Indeed, people often report that smoking marijuana seems to result in time slowing down. There is an anecdote of two hippies, high on marijuana, sitting in Golden Gate Park as a jet zooms by overhead; one of them says to the other, "Man, I thought he'd never leave."[14]

Before we go on, it should be noted that statements about time slowing down, flying by, dragging, dilating, or speeding up can be very confusing[15]—particularly when one makes the mistake of stopping to think about what such statements actually mean. Take the phrase *time flew by*. Does this imply that a clock on the wall seems to be going faster or slower? If someone reports that *time flew by*, does this mean he would accomplish less or more in a given window of objective clock time? Statements about time speeding up or flying by are inherently ambiguous. Slower and faster are relative adjectives; thus, much like saying something is to your right or left, you must provide a point of reference. When talking about temporal distortions people generally mean that external time changes in relation to a hypothetical internal clock. Let's assume that this imaginary internal clock governs our prospective judgments of time in the range of milliseconds, seconds, and minutes, and that this clock ticks ten times a second. So if it were to be sped up to twenty ticks per second in response to a threat or drugs, the result is that during a five-second period, one would be left with the impression that ten seconds had elapsed. Such a speeding up of the internal clock would generally be described as "time slowing down," "dragging," or "dilating" because we are being self-centered: we are using our internal clock as the reference and noting that external time appears to be slowing down. This account, of course, depends on the chosen clock of reference: one could also claim that time is speeding up because the internal clock is going faster than the external clock. For better or worse, however, the convention is that statements about time speeding up or slowing down refer to the apparent speed of an external clock in relation to a hypothetical internal clock—even though it is obviously the internal clock that is actually doing the slowing or speeding. It is not uncommon to see instances in the media and popular and scientific literature in which people mistakenly state that time is slowing down when they mean speeding up. Consider the cartoon in Figure 4.1: if THC, the active component of hashish and marijuana, creates the perception of external time slowing down or dragging (consistent with William James's observation and experimental evidence)—equivalent to having the internal clock speed up—then shouldn't the cowboy in the figure find the clock to be before noon? More generally, it is important to note that our feeling of how quickly or slowly time is passing is not necessarily equivalent to our explicit estimates of how many seconds or minutes have elapsed (I may estimate that I was in the dentist's chair for five minutes, but report that it felt like an hour).[16]

While we should not take the notion of a ticking and tocking internal clock literally, it provides a very useful metaphor when thinking about our perception of time. Indeed, pharmacological studies often interpret temporal distortions in the context of changes in the speed of an internal clock. For example, numerous laboratory studies with humans support the anecdotal reports that time slows down under the influence of marijuana, and these results can be interpreted as a speedup of the internal clock. In one early study, subjects were simply asked to tell the investigator when they thought 60 seconds had elapsed after being given a start signal. After being given an oral dose of THC, subjects offered significantly shorter estimates than their baseline: with THC in their system, subjects waited an average of 42 seconds before they reported that a minute had elapsed, whereas their baseline estimates were pretty close to a veridical 60 seconds.[17] It is as if their internal clock was running faster—reaching the count of 60 within just 42 objective seconds (note that the estimate is shorter because they were asked to "produce" a minute; if they had been asked to estimate the duration of an actual minute, a fast clock would result in an overestimation).

Drugs also affect the sense of time of animals. Which raises the question: how does one ask animals how much time they think has elapsed? Rats and mice can readily learn to press a lever to obtain food, and in a variant of this standard form of operant conditioning, called the *fixed-interval procedure*, a cue such as a light turning on signals the beginning of a trial (that is, $t=0$). After the start of the trial, the rat can press the lever at will, but it will only receive a food reward for the first lever press it makes after some fixed interval from the start of the trial. Rats will learn to start their lever pressing at times proportional to the fixed interval on which they are trained.

So if the fixed-interval used during training was 10 seconds, rats are more likely to press the lever at times near the fixed-interval, around 10 seconds. And rats trained on a 10-second interval will press the lever before rats trained on a 30-second interval. This is one of the many ways to demonstrate that rodents and other animals are able to prospectively keep track of time in the range of seconds. The question is what happens if after learning this task (which can take weeks of training), the rats get high? In one experiment in which rats were trained with a 30-second fixed interval, the peak time of lever pressing fell from around 34 seconds (no drug) to around 29 seconds when given THC (interestingly, in this study the rats were in a sense more accurate when on THC). That result is consistent with human reports that time slows down under the influence of cannabis because the hypothetical internal clock is running fast, although, particularly in the case of cannabinoids, such drug-induced effects on timing are not universally reproduced.[18]

The best-studied chronopharmacological effect on timing in animals involves manipulating the brain's dopamine system. Dopamine is an important neurotransmitter, and a modulator of many different brain processes. Most notably, it is damage to a cluster of dopamine-producing neurons located in the brain stem (the *substantia nigra*) that produces the characteristic tremors and motor deficits of Parkinson's disease. The Duke University psychologist Warren Meck and his colleagues have proposed that dopamine might alter the speed of the timing circuits within the brain. For example, his experiments have shown that after training rats with a fixed interval of 20 sec, the administration of the stimulant methamphetamine—which among other effects increases dopamine levels within the brain—can shift the timing of lever pressing from approximately 20 to 17 sec; but after days of repeatedly performing the task while on methamphetamines, the rats slowly readjusted their timing back to 20 sec—as if they learned to work with a chronically fast internal clock by recalibrating the number of internal ticks that correspond to 20 seconds. Furthermore, when the rats were taken off the drug, they overshot: the peak timing of the lever pressing increased to above 20 sec.[19]

These and many other pharmacological studies provide important insights into how humans and animals tell and perceive time. But making sense of these studies has been challenging and controversial. First, the results are often dependent on the nature of the task used, the interval being studied, and the details of how subjects report the perceived amount of elapsed time. Second, because virtually all drugs have multiple, interconnected neurophysiological effects, it is very difficult to determine the true cause of a change in behavior. For example, cannabinoids and dopaminergic drugs can affect levels of anxiety, memory, motor activity, and physiological states, such as hunger (which might affect the motivation of animals to perform a task). And, needless to say, these drugs can also alter the amount of attention human and animal subjects are willing to devote to the task at hand, potentially leading to a suite of alternate interpretations. Furthermore, some drugs can affect people's judgments of short intervals but not long intervals, and vice versa.[20] Overall, the scientific literature on the effect of psychoactive drugs on time perception indicates that there is no single neurotransmitter that governs our perception of time. Additionally, because the same drug can differentially affect estimates of short and long intervals, pharmacological studies provide strong evidence for the notion that there is no master internal clock that governs timing across milliseconds to hours—thus providing evidence in support of our multiple clock principle.

CAUSES OF THE SLOW-MOTION EFFECT

We have now seen that distortions of our perception of time should be considered the rule rather than the exception—thus taking away some of the mystery of the reports of temporal distortions under highly emotional or life-threatening situations. Nevertheless, reports of events unfolding in slow motion during life-threatening situations stand out because they go far beyond the notion of simply overestimating elapsed time. There are a few hypotheses that might explain why life-threatening events seem to unfold in slow motion.[21] I will mention three, which I refer to as the *overclocking*, *hypermemory*, and *metaillusion* hypotheses.

Overclocking. If the CPU on your computer operates at 2 GHz, that means it performs two billion operations in a second. This rate is controlled by the computer's "clock." The function of this clock is not to tell the time of day but to set the frequency of the operations on the CPU—in this case by sending out an electric pulse every 0.0000000005 seconds. Every gamer knows that it is possible to overclock one's computer by increasing the number of pulses the clock generates every second. The result is a computer that essentially does everything faster: it can take in and process more information in a given period of time (the downside is that the CPU might melt). Perhaps the slow-motion effect is caused by the neural equivalent of overclocking a digital computer: people's ability to react quicker and perceive events in slow motion could be explained by the brain entering an overclocked mode during life-threatening moments.

Can the brain be "overclocked"? The time it takes the brain to execute a task is determined by many different factors, including: (1) the speed at which electrical signals (the action potentials or "spikes") travel along axons; (2) the amount of time it takes the electrochemical signal at the synapse to be transmitted from the presynaptic to postsynaptic neuron—the *synaptic delay*; and (3) the amount of time it takes synaptic currents to change the voltage of the neuron enough to trigger an action potential—this is determined in part by the so-called *time constant* of a neuron. The conduction speed of axons and the synaptic delays are determined in large part by fairly rigid biophysical and biochemical events, and are unlikely to be sped up in any significant way during a flight-or-fight response. On the other hand, the time it takes for a neuron to fire in response to a barrage of inputs from its presynaptic cells could be decreased through a number of mechanisms.[22] One of the simplest ways to envision this happening is that neuromodulators (such as norepinephrine) that flood the brain and blood during fight-or-flight situations could depolarize excitatory neurons in the brain (or decrease inhibition), making it a bit easier, and a bit quicker, for them to fire. However, changes in the firing latency of neurons are unlikely to amount to speed increases of more than 10 or 20 percent. For example, there are reports that stimulants, including caffeine, can decrease how long it takes people to respond to a stimulus (their reaction time), but these decreases are generally less than 10 percent.[23] Through poorly understood mechanisms, neuromodulators can result in enhanced and sharpened attention to external events happening around us. Indeed, it is well established that performance and reaction time can be improved by attention. While such effects are likely to contribute to the impressive, on-field actions of a professional athlete, they cannot account for the striking reports of individuals who "acted with lightning quickness in accord with accurate judgment" or engaged in "a sudden review of the individual's entire past."

Reports of people performing complex, split-second lifesaving feats under heightened danger are likely to be inaccurate. And when such actions do occur, they probably come primarily from highly trained individuals: race-car drivers, fighter pilots, and extreme athletes—in other words from people whose neural circuits have benefited from thousands of hours of training. As one researcher has put it, "Skilled kayakers can adjust boat, body, and paddle so as to take exactly the one survivable line through rapids and over waterfalls. Less skilled participants perceive only confusion, and are likely to freeze, panic, or act in ways which increase danger."[24] While many researchers emphasize reports of suprahuman performance during life-threatening events, there is no shortage of accounts of people making poor decisions during these same situations. So perhaps the increased focus and honed motor skills that come with training allow professionals to quickly take action during life-threatening situations, while the rest of us—despite subjectively perceiving events in slow motion—are passively flailing and freezing in the face danger.

Hypermemory. Another possible explanation for the slow-motion effect is that it is an after-the-fact illusion—an illusion in the sense that people don't actually perceive events happening in slow motion at the time of the accident, they just think they did when they are recalling the episode. During fight-or-flight situations the brain could enhance the spatial and temporal resolution of our memories. In other words, during the threat, the perceived speed at which events transpire would be more or less normal, but during recollection memories would be much more detailed, making it seem, after the fact, that everything transpired in slow motion. In one account of the slow-motion effect a subject who was nearly killed by an oncoming train reported that "as the train went by I saw the engineer's face. It was like a movie run slowly so the frames progress with a jerky motion. That was how I saw his face."[25] But how can we determine if this account is merely generated during recall, or if it actually happened during the event? Furthermore, how do we know if the events recalled are accurate (would this individual actually be able to recognize the engineer's face)? It is well established that our memories of emotional events can be quite unreliable. We know, for instance, that there are many examples of victims of violent crimes identifying the wrong suspect during eyewitness testimonies.[26]

Nevertheless, it seems likely that some version of the hypermemory hypothesis contributes to the slow-motion effect because neuromodulators released during emotional or dangerous events can indeed enhance memory. This is thought to be one explanation of so-called *flashbulb memories*—those in which people remember where they were when they heard of a tragic event such as 9/11. Post-traumatic stress disorder is another example in which the fight-or-flight response enhances memory, in this case resulting in overpoweringly strong and maladaptive memories.[27]

The hypermemory hypothesis, of course, does not account for reports of people acting more rapidly and with more clarity than they otherwise could. Nor does it account for the often-compelling subjective sensation that the slow-motion effects occur in the moment. Here, I can't help but be influenced by my own anecdotal experience of the slow-motion effect. In an automobile accident, my car was struck from the side, spun around, and slammed into a telephone pole. My distinct feeling during the event was not only that the car was spinning slowly, but that I thought as this transpired, "Wow, time really does slow down." But to emphasize that perception is far from perfect during such moments, I do not remember reacting quickly in any way or even registering that the side airbags deployed. Nevertheless, the fact that I remember thinking time slowed down would suggest that I did perceive events unfolding in slow motion *during the accident*, and thus that the hypermemory hypothesis cannot fully account for the slow-motion effect.

Metaillusion. A shortcoming of both the overclocking and hypermemory hypotheses is that they don't take into account a fundamental observation about subjective experiences. Whether of color, sound, or the passage of time, our conscious experiences are in essence illusions, convenient running narratives of what the unconscious brain determines are the most relevant events happening in the extracranial world. This may be a strange concept—one to which we will return in chapter 12—but for now, perhaps the most compelling way to grasp what I mean by the illusory nature of subjective experiences is through the example of body awareness. One of the most intimate of all subjective experiences is that your hand is *your* hand and nobody else's. When you accidentally miss the nail and smash your finger with the hammer, you *feel* pain. Although the pain is produced within the brain, amazingly, it is not perceived as occurring within the brain; rather, it is projected out to the point in space where your finger happens to be. The illusory nature of body awareness can be unveiled by phantom-limb syndrome. Some people who have had a limb amputated continue to feel their missing limb as vividly as most of us feel our own limbs. Phantom limbs tell us that the brain is so committed to providing an illusion of ownership of the bone, muscle, and nerves that constitute our limbs that it will sometimes persevere in generating the illusion even when the limb is long gone. Thus, it is not really the phantom limb that is the illusion, but the sense of ownership of our actual limbs. Phantom-limb syndrome is a puzzling phenomenon, but focusing too much on the mystery of phantom limbs distracts from the real puzzle: how the brain creates conscious awareness of our body to begin with.[28] Similarly, focusing on the slow-motion effect in life-threatening situations distracts us from the real mystery: our "normal" perception of the passage of time.

Imagine sitting in an empty room: a movie is playing and you quickly realize the speed is all wrong. Lips are moving in slow motion and objects take too long to fall. How can you fix it? If you know nothing about how the movie is being projected, or the type of machine that is creating the projection, how could you go about trying to understand why the movie is being played at the wrong speed? After all, as far as the projector is concerned, the "correct" speed is simply one of many possible settings. Our normal sense of time is a mental construct, one that also seems to have different speed settings. The metaillusion hypothesis emphasizes that the slow-motion effect is an illusion of an illusion, so trying to explain the effect without understanding our normal subjective sense of time is like trying to fix the speed of a slow movie without knowing anything about how the correct speed is generated.

Consciousness is a delayed account of not only what is happening in the external world, but of what is happening in the unconscious brain. For example, as we will see (chapter 12), by watching the neural activity within the brain it is possible to predict when people will voluntarily decide to move their finger up to 900 milliseconds before they actually do—hundreds of milliseconds before the subjects themselves seem to be aware of having "freely" decided to move their finger. So even if danger does kick the brain into an overclocking mode—resulting in sped-up actions—consciousness might be too slow to be guiding those actions. Thus the reports of "dominant mental quickness as demonstrated by the increased speed of thoughts"[29] may just be another deception the unconscious brain imposes on the mind.

The brain can not only project the feeling of pain out into the world where our limbs are located, but when a fake arm is placed nearby your actual arm, your brain can recalibrate its projection of where in space you perceive your arm towards the position of the fake arm—as if the brain is inclined to accept the fake arm as its own (this is called the *rubber-hand* illusion). Much as the brain is able to project the feeling of the location of our limbs to different points in space, it seems to have the flexibility to differentially label events as unfolding *quickly* or *slowly*. That is, our subjective judgments as to whether time is passing fast or slowly may be dissociated from the rate the brain is processing information—or of the brain's internal clock speed.[30]

To further highlight the relevance of the metaillusion hypothesis, consider that it is not only people's sense of time that is distorted during life-threatening events. The following are three reports from a collection of over a hundred that were published in 1976.[31] The first is from a twenty-four-year-old race-car driver who was traveling at 100 miles an hour when an accident resulted in his car flipping over multiple times while flying 30 feet into the air:

> It seemed like the whole thing took forever. Everything was in slow motion and it seemed to me like I was a player on a stage and could see myself tumbling over and over in the car. It was as though I sat in the stands and saw it all happening.

A twenty-one-year-old college student involved in a serious automobile accident reported:

> *During all of this, time stood still. It seemed to take forever*
> *for everything to happen. Space too was unreal. It was all very*
> *much like sitting in a movie theater and watching it*
> *happen on the screen.*

And a soldier whose jeep was blown up by a mine during World War II recounted:

> *I had no realization of time passing, only of the moment which*
> *never altered. Neither did I have any concept of space, since my*
> *existence seemed only mental.*

Thus it appears that the perception of time is not the only mental faculty to be altered: the perception of space is altered as well. Indeed, in any other context the above reports would be called hallucinations or altered states. Perhaps the sudden flood of endogenous neurochemicals elicited by the fight-or-flight response overload the brain's circuits and induce hallucinations. So maybe the slow-motion effect is best considered another type of altered state, more divorced from reality than attached to it.

COMPRESSING TIME IN THE BRAIN

The three hypotheses described above are not mutually exclusive explanations of the slow-motion effect. I suspect that enhanced attention contributes to the swift actions of trained professionals during high-adrenalin situations, and that hypermemory likely contributes to slow-motion reports, but that ultimately the speed at which we perceive events to pass is a somewhat arbitrary setting on top of the much more mysterious illusion of consciousness.

We think of the slow-motion effect and other temporal illusions as distortions of our perception of time—more specifically of the rate at which things change—but it is not that simple. Our ability to compress and dilate time is actually a feature of the brain that we use every day.

What is the last word of the first verse of your favorite song? If you are like me you have to start from the beginning to reach the end: *How many . . . man walk . . . him a man.* But you certainly do not need to mentally replay the verse in the tempo of the original song. You can mentally replay it very quickly or slowly: fast-forwarding through the lyrics or savoring every syllable. Indeed, our ability to execute the same action at different speeds is an important feature of our motor system. I slow down my speech when speaking to babies, and speed it up when I'm running out of time during a lecture. You can tie your shoes slowly if you are teaching a child how to tie them, but do it rapidly when rushing out for a run, and you can imagine tying your shoes in your mind's eye in less time than it actually takes. Our ability to effectively speed our motor actions up or down ranges over a factor of approximately five—for example, the slowest and fastest musical tempos generally range from 40 to 200 beats per minute. But there is evidence the brain can replay events at even higher speeds.

We mentioned the place cells of the hippocampus in chapter 1. These are neurons that fire selectively when a rat is in a specific location in a room. So as a rat explores an open area, say through locations labeled 1→2→3→4→5, there will be neurons that fire at each of these locations. If we label the neurons that fire in each location as A through E, we will observe a pattern of place cells firing over time (what I will call a *neural trajectory*): A→B→C→D→E. We can think of this pattern as the neural signature of the rat's experience of running through the path. The rat may take ten seconds to run though the path, and thus the neural trajectory A→B→C→D→E will take place over this same amount of time. Now the fascinating part is that when neuroscientists record from these same cells while the animal sleeps or rests, they observe this A→B→C→D→E pattern of neural activity more than what would be expected by chance—that is, more often than if the rat had not taken the 1→2→3→4→5 path earlier in the day. One interpretation of these results is that the rat's brain is replaying the episodes it experienced earlier. But these replay events unfold on a totally different time scale; during replay the same A→B→C→D→E sequence might last only 200 milliseconds instead of 10 seconds. It is thought that these replay trajectories might contribute to the formation of memories—helping store experienced episodes within the brain's circuits (it is important to emphasize that no one is implying that the rats consciously re-experience the traversed locations during replay; they probably don't). It is also possible that this replay represents the planning of future actions. For example, when rats perform a task in which they have to stop at little "reward wells" to get a treat and then proceed to the next well, the pattern of neural activity observed during the pit stop can be used to predict where the rat will go next![32] One interpretation of this finding is that the rat's brain is planning out future actions that may take place over the course of 10 seconds in a mere fraction of a second. The compressed replay of neural trajectories, together with the fact that we can control the speed we mentally replay a song in our mind's ear, reveals that the brain can indeed process and generate temporal patterns at different speeds. It remains an open question, however, whether or not this feature of brain function is related to the subjective compression and dilation of time.

———

Our "sense" of time is not a true sense such as vision or hearing. There is no organ of time, there are no time receptors in our eyes, ears, nose, tongue, or skin. Nor could there be, as time is not a physical property akin to light or the changing pressure of air molecules. Nevertheless, the brain not only measures time, but it senses the passage of time, we seem to feel time flow. A multitude of temporal illusions reveals, however, that the accuracy of our sense of time can diverge radically from objective clock time, and much is often made of the these illusions. But the existence of temporal illusions is not surprising. Virtually all subjective experiences, including our perception of color and pain, and our body awareness, are altered by context, learning, attention, and drugs. To psychologists and neuroscientists these illusions have provided valuable insights into how the brain works, but in the end perhaps the most important lesson is that, distorted or not, all subjective experiences are in essence illusions. Thus we should not let temporal illusions distract us too much from the more fundamental puzzle: how does the brain generate a conscious feeling of the passage of time—or anything else for that matter—to begin with?

5:00 PATTERNS IN TIME

Excuse me while I kiss this guy.

—ATTRIBUTED TO JIMI HENDRIX

While you may not have given it much thought, during any conversation your brain is diligently timing the duration of each syllable, the pause between every word, as well as the overall rhythm of the stream of sounds striking your tympanic membranes.

Phonemes are the smallest unit of speech: they comprise the repertoire of sounds used in any given language (there is an approximate correspondence between phonemes and letters, but the same letter can represent different phonemes, for example, the *g* in *gun* and *gin*). Most of the time, meaning can be determined by the order of the phonemes within a phrase. In some instances, however, the same sequence of phonemes can have very different meanings, resulting in potentially ambiguous word pairs and phrases:

> *Great eyes* x *Gray ties*
> *A nice man* x *An iceman*
> *They gave her* *cat food* x *They gave her cat* *food*

These ambiguities can generally be resolved by other dimensions of speech, including the duration of syllables, intonation, stress, and the

pause between words. For example, one of the easiest ways to disambiguate the above phrases is to emphasize the pause between the
appropriate words. In the last example, a longer pause between *cat*
and *food* favors the interpretation that *her cat was given food*, whereas
a deliberate pause between *her* and *cat* suggests that *she was given cat
food*. Speech rate (speed) is also used to convey meaning and information. Consider the sentence: *The hostess greeted the girl with a smile.*
Who was smiling? Studies show that speeding up (temporally compressing) the segment "girl with a smile" biases people's interpretation towards *the girl smiled*, while slowing it down (temporal dilation)
favors the interpretation that *the hostess smiled.*[1]

A tragic consequence of such ambiguities is that there are people walking around singing the wrong lyrics to their favorite songs.
Misheard lyrics can arise because vocalists must sometimes force
phrases to fit within the structure of a song's beat (on the other
hand, some lyrics are misheard simply because vocalists are not
rewarded for elocution). There is even a name for the phenomenon of hearing multiple interpretations of a song: a *mondegreen*. A
famous mondegreen is the Jimi Hendrix line from "Purple Haze,"
excuse me while I kiss the sky, often heard as *excuse me while I kiss this
guy*. Again, as in spoken speech, such potential ambiguities relate in
part to timing, and can be resolved by emphasizing the appropriate
boundaries with pauses.

Timing is also important for the discrimination of the individual phonemes. The distinction between *b* and *p*, for example, is in
part based on the so-called *voice-onset time*: the interval between
the explosive release of air from the mouth and the vibration of the
vocal chords. If you place your fingers on your throat and say *pa*,
you can probably perceive that there is an interval between opening
your mouth and the time at which you feel you vocal chords begin to
vibrate. When you do the same thing while saying *ba*, this interval
is shorter, and probably imperceptible. The voice-onset time for *pa* is
generally above 30 milliseconds, while that of *ba* is less than 20 mil-

liseconds. The fact that we can easily hear the difference between the syllables *ba* and *pa* means that the auditory system has timing mechanisms in place to distinguish these very short intervals.

On the slightly longer scale of hundreds of milliseconds to a few seconds, timing is critical to *prosody* (the rhythm or musicality of speech). We use the intonation, timing, and rate of speech to convey emotions, sarcasm, or whether a phrase is intended as a question or not. *That was a good idea* can be either a compliment or a put-down depending on the prosody of the speaker. Studies show that changing the tempo of speech by temporally compressing or dilating sentences alters judgments about the emotional state of the speaker. In one study German speakers were asked to listen to sentences and judge the emotional state of the speaker. When participants heard sentences that were enunciated to convey sadness, subjects correctly identified the emotional state of the speaker as sad. When these sentences were sped up, participants commonly judged the speaker as being in a frightened or neutral state. Importantly, the emotions conveyed by prosody can transcend language. When the same sentences were judged by Americans who did not speak German, their judgments regarding the affect of the speaker followed the same patterns as the native Germans'. Similarly, if sentences were filtered in a manner that the words were unintelligible but preserved the overall "contour" of speech, listeners could still extract the emotional state of the speaker—think of listening to a muffled conversation through a wall: even though you cannot make out any of the words you can probably tell if they are said in anger or joy.[2]

TIMING IS FUNNY

The timing of speech is also said to be critical to comedy. I'm not sure if this observation has been carefully tested in the lab . . . or in animal studies, but the actor Sacha Baron-Cohen provided an entertaining demonstration of the adage that in comedy timing is everything in his movie *Borat: Cultural Learnings of America for Make Benefit Glorious Nation of Kazakhstan*. In one scene a humor coach is explaining the temporal intricacies of a "not" joke. Although not particularly funny, there is at least comic potential in the statement:

That suit is black. Not!

Borat tries to get this down with

> That suit is not black.
> That suit is black not.
> That suit is black.   Not!

All of which have considerably less humor potential. But why does timing contribute to a well-told joke? Humor is driven in part by surprise. For something to be funny, it should be unexpected, but must still make sense: *That suit is black, garbanzo beans* is unexpected indeed, but not particularly amusing.[3] An additional ingredient to humor may be that the unexpected must happen at the right time. The brain is continuously generating real-time predictions of *what* will happen next, and *when* it will happen: perhaps the unexpected punch line should happen within the expectation window. If the punch line comes too soon, it cannot be surprising because there is not enough time to create a prediction of what is about to happen. On the other hand, if the punch line comes too late, the listener is already mentally engaged with the next set of predictions—the element of surprise is that nothing happened, which is perplexing, but not funny.

MOTHERESE

Anybody who has ever attempted to learn a language as an adult has probably complained that native speakers talk too fast.[4] Listening to a foreign language can be like trying to recognize a face on a subway platform while you zoom by within a subway car—the brain struggles to get a grip on any specific face because they all blend together. Slowing speech down helps the novice parse strings of phonemes into isolated words.

Babies are likely to have similar struggles when learning their first language, which is probably why adults automatically slow their speech down and overstress words when talking to babies. This change in our speech patterns is referred to as infant-directed speech (or "motherese" or "parentese"). Infant-directed speech is often characterized by an increase in pitch, longer vowels, and longer pauses between words. For example, studies show that when adults are talking among themselves, the pauses between phrases are around 700 milliseconds, but when adults are speaking to babies this value increases to over a second. Studies also confirm that like adults attempting to learn a new language, infants are better able to discriminate words when they are spoken in the slowed, overemphasized, prosody of motherese.[5] Slower speech helps babies and adults parse speech: to learn where one word begins and the other ends, to prevent consecutive phonemes from interfering with each other. We will see in the next chapter that this seems to be a consequence of how the brain processes streams of information and tells time on the scale of tens to hundreds of milliseconds.

Speech is multidimensional: there are many variables that contribute to speech, including the sequence of phonemes, the interval between phonemes, the duration of syllables, the pauses between words, intonation, stress, speech speed, and the overall prosody. Many of these features require the brain of the listener to tell time. Similarly, the speaker must engage in the corresponding motor challenges necessary to generate the temporal structure of speech, including a complex sequence of tongue twists, finely timed lip movements, vocal-chord vibrations, pauses, and timed breathing. Overall, the brains of listeners and speakers alike must solve a sophisticated set of timing problems—a task that is likely beyond the capabilities of any simple clock-like device.

MORSE CODE

We have seen that humans and other animals engage in a wide range of temporal tasks, timing the delays it takes sound to arrive from one ear to another, the duration of red lights, or the rotation of the Earth around its axis. These tasks rely on the timing of isolated intervals or durations; the temporal equivalent of judging the length of an object. In contrast, speech and music recognition require determining the temporal structure of complex temporal patterns: of putting together many temporal pieces to make out the whole.

Time is to speech and music recognition as space is to visual object recognition. We can think of recognizing a face in a drawing as a spatial problem—that is, the relevant information is contained in the spatial relationships between all the elements of the drawing. It is also a hierarchical problem: low-level information (lines and curves) must be integrated into a unified image. A circle is a circle, but two side-by-side pairs of concentric circles become eyes; place those in a larger circle and you have a face, and so forth until we have a crowd of people within a scene. Speech and music are the temporal equivalent of recognizing a visual scene: they require solving a hierarchy of embedded temporal problems.[6] Speech requires tracking the temporal features of progressively longer elements: phonemes, syllables, words, phrases, and sentences. In some ways recognizing a hierarchy of temporal patterns is more challenging, because it requires some sort of memory of the past. All the features of a drawing are simultaneously present on a static piece of paper, but the relevant features of speech or music require integration across time; that is, each feature must be interpreted in the context of elements that have already faded into the past.

Morse code provides perhaps the best example of just how sophisticated the brain's ability to process temporal patterns is. Speech and music rely on information encoded in the temporal structure of sounds, but there is also a vast amount of information conveyed in the pitch of sounds. We can think of pitch as spatial information, a bit like the orientation of a line on a piece of paper. This can be a bit confusing because pitch refers to the perception of the frequency of sounds, and frequency is an inherently temporal property measured in cycles per second—that is, the interval between the repetition of cycles of sound-wave vibrations. The frequency of sounds, however, are represented spatially by the auditory sensory cells (*hair cells*) along the length of the cochlea. So as far as the central nervous system is concerned, discriminating the pitch of sounds is essentially a spatial task—akin to the differences in the location of the keys on a piano. Morse code is independent of pitch or spatial information of any sort—in Morse code timing *is* everything.

There are two fundamental elements of Morse code: *dots* and *dashes*. The only difference between these symbols is their duration, thus Morse code only requires a single communication channel, such as a tone or light going on and off in some complex temporal pattern. This simplicity makes the code easy to transmit. Messages can even be sent with short and long eye blinks. This was famously done by the American admiral Jeremiah Denton during the Vietnam War. As a prisoner of war he was interviewed for propaganda purposes, and during the televised interview he answered a question by saying "I get adequate food and adequate clothing and medical care when I require it." But as spoke he blinked: T O R T U R E.[7]

The duration of a dot and a dash depends on the overall Morse code rate, measured in words per minute. At the rate of 10 words per minute, the duration of each dot and dash is 120 and 360 milliseconds respectively. But there is information encoded in the pauses as well: the pause between each letter is 360 ms (3x the duration of a dot), and the pause between each word is 840 ms (7x the duration of each dot). The pattern

. ▬ ▬   . . . .   . ▬ ▬   ▬   . .   . . .   ▬   . .   ▬ ▬   .

reads *what is time*. The longer pauses represent the break between words. All the information is contained in the duration of tones, the interval between them, and their overall global structure. But like speech, there is also a prosody to Morse code, and experts seem to be able to use slight variations in the timing to identify the speaker through his or her "accent." To the untrained ear, listening to a long Morse code message is much like listening to a foreign language: it is impossible to hear when one letter ends and the next begins. Each incoming tone just piles onto the last, making it impossible to discriminate between the sequences:

· · ·   · · · ·   · (she)

and

· · · ·   · ·   · · · (his)

Of course, an expert no more needs to consciously count these dots and think about when one letter ends and the next begins than you need to stop and think about whether you heard the letter *t* when distinguishing between the words *neuron* and *neutron*.

So how does one become a Morse code expert? Slowly. One does not start learning Morse code at 20 words per minute: people start at slow rates and work their way up. One recommended method uses so-called Farnsworth timing: the letters are transmitted at normal speed, but the pauses between the letters and words are emphasized by lengthening them. This allows students to learn the letters as a single "temporal object," while emphasizing the boundaries between each letter and word so that they interfere less with each other.[8] In other words, people learn Morse code by starting with Morse code motherese.

LEARNING TO TELL TIME

Even for those uninitiated in Morse code, it is relatively easy to discriminate a single 120 ms dot from a 360 ms dash. Similarly, in the context of music, it is easy to discriminate a single 250 ms note from a 500 ms note (an eighth and quarter note at 120 beats per minute respectively). But how are these simple forms of temporal discrimination achieved by the brain? Does timing get better with practice? Answering these questions provides important insights into how the brain tells time.

One might guess that the brain uses some sort of all-purpose neural stopwatch to time all durations in the range of a few milliseconds up to a second or so. On the other hand, we might speculate that the brain has a multitude of different neurons or circuits each specialized for detecting a given interval, akin to having a collection of hourglasses—one for each possible interval. To attempt to distinguish between these hypotheses we can ask if, and how, people's ability to discriminate intervals improves with practice.

Although interval discrimination has been studied since the late nineteenth century, it was only in the 1990s that the question of whether timing improves with practice was conclusively answered. One of the first studies to systematically address this question was performed at the University of California, San Francisco, by Beverly Wright, myself, and our colleagues Henry Mahncke and Michael Merzenich. We used a standard interval-discrimination task in which subjects heard two different intervals and were asked to decide whether the first or the second interval was the longer. In this task, each interval was bounded by two brief tones (15 ms each). Thus, the first interval might consist of the two tones separated by a *standard interval* of 100 ms, while the second, so-called *comparison interval*, might consist of two tones separated by 125 ms (Figure 5.1). The difference between the standard and comparison interval, in this case 25 ms, is referred to as delta-t (Δt). If, presented with intervals of 100 and 125 milliseconds, the participant always correctly identifies the longer interval, we can conclude that her internal timer has a resolution of better (less) than 25 milliseconds.

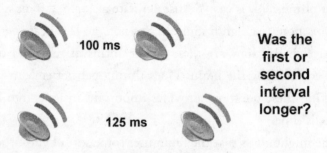

100 ms

Was the first or second interval longer?

125 ms

Figure 5.1: Interval Discrimination Task.

By varying the value Δt it is possible to estimate the precision of the brain's timers. We first estimated the threshold of subjects on standard intervals of 50, 100, 200, and 500 ms. The first thing to note is that these thresholds were very different for each standard interval, this is a general property of how humans discriminate stimuli of different magnitudes. You can probably tell the difference between two objects that weigh 100 and 125 grams, but not between objects that weigh 1,000 and 1,025 grams. Generally speaking, what matters is not the absolute difference between the two stimuli, but the relative ratio between them. The interval discrimination thresholds were around 15 to 25 percent. For example, for the 100 ms standard interval the average *interval discrimination threshold* was 24 ms, meaning that on average people could reliably discriminate 100 from 124 ms. After getting this baseline data on the first day of the study, the subjects underwent a ten-day training period in which they practiced discriminating the 100 ms intervals for an hour a day. After this practice period, the subjects' timing did indeed improve, the average threshold for the 100 ms standard fell from 24 to 10 ms. This suggests that practice does somehow improve the quality of the timers within our brain. But humans are complicated creatures; perhaps practice did not improve timing per se, but appeared to because over time the volunteers were better able to focus on the task. Fortunately the answer to a second, and more interesting question invalidated this interpretation.

Given that people improved at the trained 100 ms interval, do they improve at the other intervals as well? Note that if we believe the brain has some sort of generic neural stopwatch responsible for timing all intervals between 50 and 500 ms, and that this stopwatch somehow becomes more accurate with practice, we would expect that timing of all intervals would improve, even though the volunteers only practiced on the 100 ms. In contrast, if the brain uses specialized timers, then we would predict that the improvements on the 100 ms standard interval would not generalize to the other intervals. This was indeed the case. Although ten days of practice on the 100 ms intervals dramatically improved people's ability to discriminate intervals around 100 ms, it did not improve the interval discrimination thresholds for the 50, 200, or 500 ms intervals at all.[9] If learning on the 100 ms interval came from improved focus, then subjects would likely improve on all intervals, but we did not observe that. More importantly, this result, which has since been replicated in numerous other studies,[10] suggests that however the brain is telling time in the subsecond range, it does not seem to be via any sort of master stopwatch mechanism that can time any and all intervals.

It stands to reason that if timing improves with practice, people in professions that require accurate timing—such as musicians—should be better than average. One early study that addressed this question was performed by Richard Ivry and colleagues, then at the University of Oregon. They asked pianists and nonpianists to simply press a button in synch with a series of tones presented every 400 ms, and then to continue tapping away with the same timing after these pacing tones ceased. The intertap intervals produced by the musicians were significantly less variable (more consistent) than those produced by nonmusicians. Similarly, pianists were also better at an interval discrimination task with a standard interval of 400 ms.[11] Another study confirmed that interval-discrimination thresholds of musicians were significantly lower for standard intervals of 50 and 1,000 ms. But even among musicians there are significant timing differences. Drummers, for example, have been shown to discriminate 1 sec intervals better than string musicians.[12] Overall, studies reveal that musicians generally perform at least 20 percent better than nonmusicians in a variety of temporal tasks.

KEEPING THE BEAT

Music, in one form or another, is universally present across human cultures. A key ingredient of music is its beat: the periodic pacing that serves as the foundation of a song's rhythm. Our natural tendency to gravitate towards the beat of the song by tapping, or bobbing our head, is one more example that the human brain is a prediction machine. You do not tap your foot in *response* to each beat—which is often marked by the strike of a drum—rather, your brain is looking a few hundred milliseconds into the future to predict when the next beat will occur, and synchronizes your movements to match it. Synchronizing our movements with the beat of a song is so easy that sometimes it is easier to tap along than suppress the allure of the rhythm. Yet most animals do not possess the simple ability to keep a beat.

It is not only that animals don't share our musical proclivities; rather, they seem to lack the sensory-motor skills necessary to synchronize their movements with a periodic stimulus. At this point any YouTube connoisseur will object, and point out the abundance of cute pets happily bobbing along with the beat of some pop song. Some of these videos are probably lucky Clever Hansian effects: animals that have learned to follow clues of their owners, like the famous horse who performed arithmetic by following his owner's involuntary body cues. Other videos however—particularly the bird clips—may be the real thing.

Scientists are certainly not above recruiting their subjects from YouTube videos. A study performed by the psychologist Aniruddh Patel and his colleagues enlisted the YouTube star Snowball, a cute white cockatoo.[13] In one of his videos Snowball engages in a series of body and head-bob movements—which can only be described as dancing—to the tune of "Everybody" by the Backstreet Boys. To determine if Snowball was actually following the beat, as opposed to having memorized a series of fixed moves, the investigators slowed or sped up the song, and determined where Snowball's movements fell in relation to the beat of the song; if the head is always in more or less the same extended position at each beat, we can say that his movements synchronized with the beat. Snowball's moves were clearly in synch with the beat over a range of tempos, meaning that he was anticipating the beats—although he did seem to prefer dancing to the faster beats.[14] But birds are the exceptions. Monkeys can learn to reproduce a single interval marked by two auditory tones, but they struggle to perform a simple synchronization task. One study reported that even after a year of training, rhesus monkeys were not able to press a button in synchrony with periodically presented tones, although they were able to consistently tap the button slightly after each tone.[15]

So why is the seemingly trivial task of keeping a beat so challenging for our fellow primates, yet not for some birds? One potential answer to this question is referred to as the *vocal learning hypothesis*. Most mammals, including monkeys, dogs, and cats, communicate with each other through cries, howls, growls, barks, or meows, but these behaviors are innate, and reflect a very simple and limited set of "words"—a dog, for example, does not need to learn that a growl does not mean "welcome, please come closer." Relatively few animals learn to produce vocalizations as a result of experience and social interactions. In addition to humans, species that are capable of vocal learning include some birds, whales, and elephants. Parrots are the most obvious example, as they can learn to reproduce the sounds made by other birds, or to imitate a limited repertoire of words from the pirate vocabulary.

Vocal learning requires that the brain listen to sounds and then figure out how to reproduce those same sounds using the vocal chords and oral muscles. This task clearly requires significant cooperation between the brain's auditory and motor centers. Similarly, the ability to move in synchrony with a periodic auditory stimulus also requires a tight cooperation between the auditory and motor systems. It has been proposed that the same brain wiring that allows animals to learn vocal communication also underlies the apparently much simpler act of following the beat of a song.[16]

Speech and music are active endeavors that require the brain to create a running expectation of what will happen next. Music, in particular, is about being led to anticipate a particular note at a particular time; whether that expectation is satisfied or violated depends on the composer's intent.[17] It is thus not surprising that the ability to follow a beat is a minimal requirement for music appreciation, as tapping in synchrony with a periodic stimulus is one of the most basic measures of prediction and expectation.

SONGBIRDS

Birds not only dance, some can sing. At least *we* call it singing—they are communicating with one another. There are numerous parallels between song learning in songbirds and human speech. These similarities have made songbirds an important species for studying learning, communication, language, and timing.[18] Male zebra finches vocalize elaborate songs as part of their courtship behavior. Young males learn these songs from adult males—or even by listening to an audio recording of another male singing. Much like human speech, there is a critical developmental window over which vocal learning must occur. If that early developmental window is missed, songbirds will never learn to produce a normal adult song—a male that has never heard another male's song will sing, but the quality of its song is unlikely to seduce any females.

As in speech and music, there is a temporal hierarchy of elements within a bird's song. Notes are combined to form syllables, and a sequence of syllables forms the phrases within a song. A given syllable can last up to a few hundred milliseconds, the pauses between syllables generally last less than 100 ms, and an entire song can carry on for seconds. The brains of male and female zebra finches are quite different: the males have a number of brain areas that are critical to song learning and production (the females do not sing). One such nucleus is referred to as HVC (which stands for nothing—don't ask). This nucleus is at least in part responsible for the timing of a zebra finch's song. Neurons in HVC fire at specific times during the song—for example one neuron might fire 100 milliseconds into a phrase, while another may fire around 500 milliseconds in.[19] We can think of these neurons as forming a neuronal chain in which neuron A activates B which activates C... (Figure 5.2). The result is that once neuron A fires we have a domino effect of neuronal activation $A{\rightarrow}B{\rightarrow}C{\rightarrow}D{\rightarrow}E$ (in reality it is best to think of each link in this chain as a group of neurons rather than a single neuron). This is like using a chain of falling dominos as a timer: if the dominos are arranged in the same position time and time again, and each domino takes 100 milliseconds to fall, then we know that approximately 500 milliseconds have elapsed when the fifth domino falls, and 1 second has passed when the tenth domino falls, ... Similarly, as we will examine in more detail in the next chapter, one theory is that in some cases the brain tells time by determining which neurons in a chain are currently active. Neurons in HVC indeed seem to use such a mechanism to control the timing of the notes of a bird's song. But one of the chronic challenges in neuroscience is to distinguish between correlation and causation—just because neurons in HVC seem to be responsible for singing does not mean they actually are. As one approach toward addressing this correlation-versus-causation question, the neuroscientists Michael Long and Michale Fee, both at MIT at the time, reasoned that if HVC neurons were causing the timing of the song, then if the pattern of activity of those neurons was slowed down, the birds should sing in slow motion.[20]

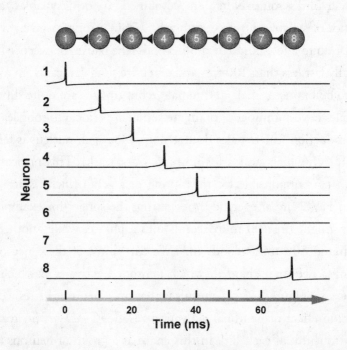

Figure 5.2: Synfire Chain. In a synfire chain model individual neurons (or groups of neurons) are connected in a feed-forward fashion. Activity—action potentials represented by "spikes" in voltage—propagates throughout the network much like a pattern of falling dominos. Time from the activation of the first neuron in the chain can be encoded by which neuron is currently active.

Slowing the activity of a group of neurons is a delicate endeavor, but it is possible to do by manipulating the local temperature of a targeted area of the brain. Cooling biological tissue generally slows its metabolism and rate of activity. The same is true of neurons. For example, in ectothermic ("cold-blooded") animals the speed at which an action potential travels along an axon and even the duration of the action potential itself can depend on external temperature (this is one reason endothermic animals generally have quicker reflexes than ectothermic animals). To lower the temperature in HVC, Long and Fee used a tiny cooling element that could be inserted into the brain of birds. This allowed them to decrease the temperature of HVC five or six degrees centigrade below body temperature as male birds sang (males are generally coaxed to sing by placing a female in a neighboring cage). The results were clear-cut. As far as song production was concerned, cooling HVC slowed the rate of singing. Notably, this slowing of the song's tempo was uniform across the entire song; that is, the notes, syllables, pauses, and overall phrase length were stretched by the same amount, up to 40 percent. As an important control experiment, the investigators also cooled a motor nucleus important for song production—an area that receives input from HVC. Cooling this area did not significantly affect the song timing, implying that the effect is caused by slowing down the activity patterns (the neural dynamics) *within* neurons in HVC, and that in this instance the motor area operates more or less as a slave to HVC.

There are many unanswered questions regarding the timing of the songs of birds, but these experiments provide one piece of evidence that a single nucleus within the brain may contribute to, if not govern, the timing and temporal structure of a complex behavior.

THE NEUROANATOMY OF TIME

Electrophysiological studies in animals and imaging studies in humans do not provide converging evidence that there is any master circuit within the brain responsible for telling time on the scale of hundreds of milliseconds to a few seconds. Much to the contrary: it is increasingly difficult to find some area in the brain that has not been implicated in timing of one sort or the other.[21] It is clear that any strong version of a master clock strategy is incorrect, which is not to say that specific areas in the brain are not responsible for some specific forms of timing. In songbirds, HVC indeed seems to be critical for song timing. As we will see in the next chapter, in mammals the cerebellum is important for some forms of motor timing. Furthermore, studies have consistently implicated some areas of the human brain in the discrimination of intervals. These areas include the basal ganglia (a group of brain nuclei located below the cortex) and the supplementary motor area (an area adjacent to the motor cortex that contributes to movement).[22] Yet it is too early to say if these areas are actually telling time or are rather reporting the time—that is, whether they are the quartz crystal or the digital display of a wristwatch. Additionally, these studies don't reveal much about *how* any circuit within the brain tells time—that is, the neural mechanisms of timing.

Theoretical and experimental research from my and a number of other laboratories suggests that while specific circuits within the brain are responsible for certain types of timing, most neural circuits are intrinsically able to tell time if needed. Depending on the characteristics of the task—for example, sensory versus motor timing, interval versus pattern timing, or subsecond versus second timing—specific neural circuits may be primarily responsible for timing. Thus auditory circuits might be partially responsible for distinguishing a quarter musical note from an eighth note; visual circuits might contribute to the discrimination of a visually presented Morse code dot or dash; motor circuits may be responsible for tapping out a SOS signal in Morse code; and the basal ganglia may contribute to our ability to anticipate when a traffic light should change.

This view that timing is a general computation that most neural circuits can perform—to one degree or another—has led my lab to ask: *can an isolated piece of the cortex, kept in culture in a dish, tell time?*

Much as it is possible to keep blood cells or heart or liver tissue alive in vitro ("in a dish"), neuroscientists have long been able to culture pieces of cortex obtained from rats or mice. These in vitro cortical circuits can contain tens of thousands of neurons, and be kept alive for weeks or months. Typically, these circuits sit in an incubator, deprived of any interaction with the external world. Hope Johnson and Anu Goel in my lab asked what would happen if these circuits were exposed to some sort of temporal pattern. Would the circuits change or adapt in any way? Could the circuits in a sense "learn" a specific interval? In one series of experiments brain slices from the auditory cortex of rats were electrically stimulated with intervals of 100, 250, or 500 ms for a few hours.[23] Normally the brain receives information through its sensory organs, the neurons sitting in a dish do not have any way to receive signals from the outside world. To provide the in vitro circuits with sensory experience of sorts, metal microelectrodes were used to provide a brief shock to the tissue, which caused a small percentage of the neurons to fire. At intervals of either 100, 250, or 500 ms Anu Goel applied a second stimulus, this one in the form of a pulse of light that also caused a subset of the neurons to fire. Normally, of course, neurons do not respond to light (with the exception of the photoreceptors in our eyes), as they lack the pigments that detect light. However, through so-called opto-genetic methods, it is possible to coax the neurons in a dish to fire in response to light by transfecting them with a gene that codes for a protein that is light sensitive.[24] Thus these cortical circuits now had very limited contact with the outside world: all they experienced was one of three different temporal intervals. The question was whether their experience affected in any way the behavior of these circuits. Naive slices will often respond to a brief electrical pulse with a burst of network activity that lasts up to a few hundred milliseconds. This occurs because the neurons directly activated by the shock will activate other neurons, which in turn might further activate others— the activity "reverberates" for a few hundred milliseconds until the activity dies out. This activity is a signature of the internal dynamics of the network. Depending on the interval used to train the slices, the internal dynamics of the network exhibited different signatures. When trained on a short interval the activity was short lasting; when the slices were trained on intervals of 250 ms or 500 ms, the average duration of the evoked network activity lasted progressively longer. Thus, not only was the internal dynamics of the slices altered by their experience, but the temporal profile of the dynamics adapted to the trained interval. Independent work in the laboratory of Marshall Shuler at Johns Hopkins University also observed a form of interval learning in in vitro cortical slices from the visual cortex.[25] One interpretation of these results is that even cortical circuits in a dish are able to, in a sense, learn to tell time.

These in vitro studies provide strong support for the notion that we should view timing in the range of hundreds of milliseconds not as a computation performed by specialized circuits, but as an intrinsic property of neural circuits.

People often ask if there are any neurological disorders that result in the loss of the ability to tell time. The answer to this question depends on what time scale we are referring to. We have seen that the amnesic patient Clive Wearing certainly lost track of the passage of time of minutes, which is why he appeared to be stuck in a perpetual loop of believing that he just woke up. This makes sense; even if we have a working clock hanging on the wall we would not be able to use it to determine that an hour has elapsed if we can't remember when the task started. Because Wearing's ability to play music, understand speech, and talk were all intact, his timing on the order of hundreds of milliseconds is clearly intact. And presumably, like the famous amnesic patient H. M., he can accurately reproduce intervals of a few seconds.[26] Patients like Wearing and H. M. have a severe impairment in their ability to form new long-term memories (more specifically, memories about facts or episodes of their lives). Such deficits led to fundamental insights into how the brain stores memories. So it is natural to also ask if specific disorders abolish people's ability to tell time on the scale of around a second. The answer is no. There are no known neurological conditions that result in people losing their ability to appreciate the rhythm of music, *and* reproduce intervals in the range of seconds, *and* learn to blink at the right time in response to a tone. Nor should we expect there to be because different temporal problems are solved by different circuits within the brain.

———

Objects are physical entities that can be seen or touched, but the brain itself never sees or touches anything directly. All of its knowledge of the outside world arrives through patterns of action potentials created at one of five sensory organs. From these patterns the brain learns to identify actual physical entities, such as light sabers or papayas. Visually speaking, such objects are in a sense independent of time; they can be identified from a snapshot. But much of what the brain identifies in the external world is inherently temporal in nature: a wave of the hand, a ripple on a pond, the letter · · · (*s* in Morse code), a catchy tune, the swing of the bat, and the spoken words *swing* and *bat*. These are all temporal "objects." To detect, and represent, such events, the brain needs to be sensitive to temporal order and timing.

The voice onset time of a phoneme, the duration of a musical note, or the difference between a Morse code dot and dash all require low-level timing—they are the trees within the forest. But speech, music, and Morse code are characterized by the larger temporal landscape. Seeing this landscape is only possible on the time scale of tens of milliseconds to a few seconds. Speech and music do not exist outside this window. Slow music down or speed it up too much and it ceases to be music. Speak too fast and the phonemes fuse with each other; slow speech too much and the phonemes are not recognizable, and we begin to lose track of the previous phonemes and words in a sentence.

Subsecond timing is the Goldilocks zone for temporal processing; it is where we are able to make sense of both the forest and the trees.

Without the ability to parse complex temporal patterns we would be unable to engage in two signature abilities of the human species: speech and music. But how does the brain solve the complex timing problems inherent to speech or music? How does the brain measure the duration of a syllable or determine the tempo of a song?

6:00 TIME, NEURAL DYNAMICS, AND CHAOS

> What is a clock? The primitive subjective feeling of time flow
> enables us to order our impressions, to judge that one event
> takes place earlier, another later. But to show that the time
> interval between two events is 10 seconds, a clock is needed.
> By the use of a clock the time concept becomes objective. Any
> physical phenomenon may be used as a clock, provided
> it can be exactly repeated as many times as desired.
>
> —EINSTEIN AND INFELD[1]

Man-made clocks rely on an embarrassingly simple principle: counting the number of cycles of an oscillator. The sophistication of the oscillator varies immensely—the swing of a pendulum, the vibration of a quartz crystal, or the cycles of the "vibrations" of electromagnetic radiation—but in the end man-made clocks are simply about counting the ticks and tocks of some periodic process. Given how absurdly successful this time-keeping strategy has been, it is tempting to assume that the brain relies on similar principles to tell time.

Perhaps too tempting.

The most influential theory of how the brain tells time on the scale of milliseconds to seconds is called the *internal clock model*, first outlined in the early sixties.[2] As the name implies, this model relies on a principle similar to that of man-made clocks: a neuron or a group of neurons would beat at some fixed frequency, while another group would count the number of these ticks. This seems like a sensible proposal, particularly once we learn that many neurons do indeed oscillate, which is to say, they can fire over and over again at a fairly fixed interval. Indeed, oscillating brain waves, breathing, walking, and the heartbeat are all examples of highly rhythmic biological phenomena that rely on biological oscillators with periods ranging from tens of milliseconds to a few seconds.

But man-made clocks require more than a good oscillator; they require a mechanism to count each oscillation—the gears of a mechanical clock or the digital circuits of a quartz watch perform this function. And therein lies a problem: whereas neurons can be gifted oscillators, counting is not their forte.

SUPRA- AND INFRAPERIOD TIMING

Perhaps you are thinking, *Wait a minute, we've already seen that circadian clocks rely on a biological oscillator—one that depends on a transcription/translation autoregulatory feedback loop.* Furthermore, as just mentioned, the timing of our breathing, heartbeat, or walking does indeed rely on biological oscillators. Thus the body does use the same principle as man-made clocks to tell time! This line of reasoning is only partially correct because there is an important difference between these examples and man-made clocks. The relevant intervals being timed in these biological examples are *less than or equal to* the period of the oscillator, whereas the opposite is true of man-made clocks; man-made clocks can only tell time on scales that are *above* the period of their time base. Biological oscillators are generally used to time events of durations less than their period—*infraperiod timing*—whereas man-made clocks measure elapsed time for durations above their period—*supraperiod timing*.

The molecular machinery that comprises the circadian clocks discussed in chapter 3 has a period of around 24 hours. The concentration of circadian proteins, such as Period, provides information as to the phase within this 24-hour cycle—for example, whether it is morning, afternoon, or night. But the circadian clock within your suprachiasmatic nucleus has no idea now many days have passed! Each day is a complete reset: like a solitary pendulum that is not hooked up to any gears, there is no record or memory of how many cycles have transpired. Similarly, the neural oscillators that underlie breathing ensure that we breathe at more or less the same frequency, say 0.25 Hertz (a period of 4 seconds). Each breathing cycle requires a number of well-timed motor events, including the coordinated control of inspiration and expiration. Thus, neural centers controlling breathing can be said to tell time on a scale of less then 4 seconds, but again each period is essentially a complete reset.[3] The neural circuits controlling breathing have no idea if they've generated one thousand, one million, or one million and one breathing cycles.

There are networks of neurons that integrate information across time ("count"), but they lack the digital precision and memory span of the gears of a pendulum clock or of the digital circuits that count the ticks of a quartz or atomic clock. Well-trained humans can distinguish a 100 ms interval from a 105 ms interval. But to detect this 5 ms difference using a supraperiod timing mechanism, the time base would have to oscillate at 200 Hz, and the accumulator circuit would have to distinguish between 20 and 21 ticks, requirements that are difficult to meet given the temporal properties and precision of neurons. Consistent with this observation, there is little experimental support for the internal clock model of timing.

The lack of empirical support for the internal clock model, does not necessarily mean that the brain's oscillators might not be involved in supraperiod timing. For example, it has been proposed that some forms of timing could rely on a collection of neural oscillators, each ticking at a different frequency. When this array of oscillators is set into motion, different subpopulations of neurons will form *beats*—moments in which some of the oscillators are temporarily aligned, followed by moments in which most of the oscillators are out of phase with each other. Computational models have shown that by detecting these beats, neural oscillator models could be used for discriminating intervals below the period of any one of the oscillators.[4]

Yet we have seen that timing on the scale of hundreds of milliseconds to a few seconds is very special. Within this range timing is about more than diligently measuring the interval between events, it is about context, temporal hierarchies, and patterns in time. Within this Goldilocks zone of pattern timing we can extract both the duration and interval of events, as well as the overall temporal structure of sequences of phonemes, musical notes, and the dots and dashes of Morse code. So it seems that we might want to look beyond conventional oscillator-based mechanisms to account for timing on the scale of milliseconds and seconds.

RIPPLES

Consider the patterns of ripples created by two raindrops falling in a pond, as shown in Figure 6.1. Which of the raindrops fell first? One of the goals of this chapter is to demonstrate—as Einstein and his colleague Leopold Infeld remind us in the epigraph of this chapter—that in principle, any physical phenomenon that can be repeated in a reproducible manner can be used to tell time.

Assuming each raindrop hits the surface of the water with more or less the same momentum, each creates a similar pattern of outwardly expanding concentric waves. These ripples are an example of a *spatiotemporal pattern*—a spatial pattern that is changing over time. A snapshot of the spatial pattern at any moment in time not only makes it obvious which raindrop fell first but, with a bit of math, also allows us to estimate the interval between the raindrops.

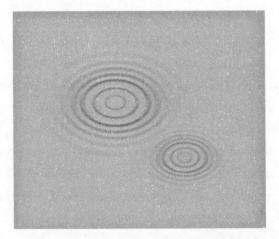

Figure 6.1: Ripples. Time is naturally encoded in
the state of dynamical systems. Here it is clear
which raindrop fell first, and it would be possible to
estimate the interval between the raindrops.

Let's consider one more simple example of how a physical system
that changes in time could potentially be used to tell time. Envision
a child sliding down a waterslide: if she goes down starting from the
same initial position every time, she will take approximately the same
amount of time to reach the bottom. We could thus mark the slide
with lines representing one-second intervals, which would have small
spacings at the top, and larger spacings at the bottom to account for
her increasing speed as she moves down the slide. Thus, as the child
crosses each line we could call out how much time has elapsed since
she started.

Our child-on-a-slide timer is driven by gravity, much like a water
clock or an hourglass. Such timers might not seem particularly precise, but consider that the top eight men in the downhill skiing com-
petition of the 2014 Winter Olympics made it down within a half a
second of each other. The top eight times ranged from 2:06:23 to
2:06:75, an accuracy of less than 0.4 percent—better than any clock
invented before Huygens's pendulum clocks.

SHORT-TERM SYNAPTIC PLASTICITY

Most physical systems—a downhill skier, a ball rolling down a ramp, the biochemical reactions within a cell, or the ripples on a pond—are processes that unfold in time in a manner determined by the laws of physics—that is, they are dynamical systems that can in principle be used to tell time. The brain is the most complex dynamical system in the known universe, thus it seems natural that the brain might tap into its intrinsic dynamics to tell time. Indeed, every time a neuron fires it undergoes a series of reproducible changes—much like those produced by the raindrops falling into a pond.

We have seen in chapter 2 that neurons are connected by synapses, and that the strength of a synapse determines the influence the presynaptic neuron will have on the postsynaptic neuron. Furthermore, the strength of these synapses can change—a weak synapse can become strong—and the process of synaptic plasticity is one way the brain learns and stores information.

To simplify our lives, neuroscientists often pretend that in the absence of learning the strength of the synapse stays more or less constant. However, most synapses become temporarily stronger or weaker every time they are used, that is, after each presynaptic spike. These use-dependent changes in synaptic strength are referred to as *short-term synaptic plasticity*, and occur over the range of tens of milliseconds to a few seconds.[5] Some cortical synapses exhibit short-term facilitation; for example, if the presynaptic neuron generates two consecutive spikes 100 milliseconds apart, the second spike will produce a larger change in the voltage of the postsynaptic neuron than the first (Figure 6.2)—that is, the message being sent from the presynaptic to postsynaptic neuron becomes "louder." Most cortical synapses, however, exhibit short-term depression—that is, the second of a pair of spikes 100 milliseconds apart produces a smaller voltage deflection in the postsynaptic neuron. Either way the magnitude of the voltage deflection depends on the interval between a pair of spikes; generally the effect is maximal at intervals below 100 ms, and fades away after a few hundred milliseconds. What this all means is that much as the diameter of the leading wave of a ripple on the pond contains information about how much time has elapsed since a raindrop fell, the strength of a synapse at any given moment contains temporal information about how long ago that synapse was last used.

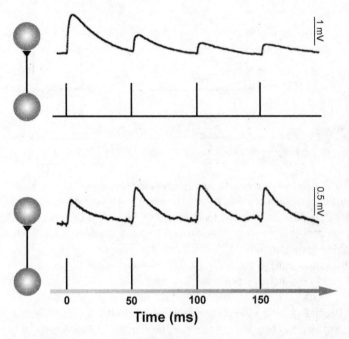

Figure 6.2: Short-term synaptic plasticity. On the time scale of milliseconds the strength of a synapse can undergo short-term depression (above) or short-term synaptic facilitation (below). (Traces from Reyes and Sakmann, 1999)

I have proposed that short-term synaptic plasticity and other time-dependent neuronal properties may contribute to the brain's ability to tell time on the order of hundreds of milliseconds.[6] Consider the simplest neural circuit possible: two neurons connected by a single synapse (Figure 6.3). Let's assume that the presynaptic neuron fires at one of three different temporal patterns: two spikes separated by an interval of either 50, 100, or 200 milliseconds. We can think of

A

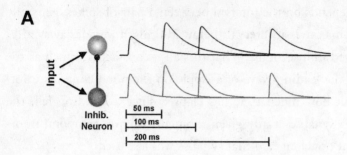

B

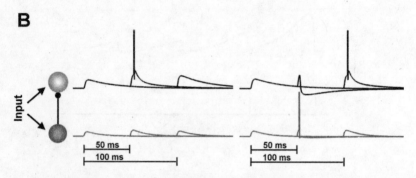

Figure 6.3: Interval selectivity based on short-term synaptic plasticity.
A. In this simulation of a simple neural circuit, a single input neuron contacts an excitatory (top) and inhibitory (bottom) neuron. The traces capture the voltage deflections in response to three different intervals: the input neuron fires two spikes separated by 50, 100, or 200 ms. The synapses from the input onto both neurons undergo short-term facilitation—for example, the amplitude of the voltage signal in response to the second spike of a 50 ms stimulus is larger than the voltage deflection caused by the first spike.
B. Depending on the strength of the synapses from the input to the excitatory and inhibitory neurons, the excitatory neuron can selectively respond to a 50 (left) or 100 ms (right) interval—thus the excitatory neuron in this simple circuit can, in a sense, tell time.

these intervals as temporal stimuli—indeed, some animals communicate with "clicks," brief bursts of sounds in which the interval between the clicks conveys information. For each of these three intervals, the voltage change produced by the first spike will be the same—and we will assume that the "strength" of this spike is 1 millivolt. Because of short-term synaptic plasticity the strength of that same synapse will be different at the time of the second spike. Specifically, the change in voltage produced by the second spike of a 50 ms interval may be 1.5 millivolts, while the strength of the second spike of 100 and 200 ms might be 1.25 and 1.1 millivolts respectively. If we construct the properties of our circuit so that the postsynaptic neuron only fires when it receives an input of at least 1.5 millivolts, then we would have constructed a timer of sorts—a neuron that only fires when it receives two inputs 50 milliseconds apart.

Neurocomputational models have shown that simple circuits composed of excitatory and inhibitory neurons that exhibit short-term synaptic plasticity can respond selectively to a range of different temporal intervals—for example, to a 100 ms interval, but not a 50 or 200 ms interval.[7] Such interval-tuned neurons could potentially be used to detect the voice-onset time of phonemes, the interval between Morse code symbols, or between musical notes.

Research has found neurons that respond selectively to different intervals in the brains of many different animals, from crickets to electric fish to rats. Exactly how this temporal specificity arises is not fully understood, but some studies indicate that short-term synaptic plasticity is at least in partially responsible.[8]

STATE-DEPENDENT NETWORKS

Our trivialized two-neuron circuit above is a bit of an insult to the brain's actual circuits. A cubic millimeter of cortex can contain a hundred thousand neurons and hundreds of millions of synapses.[9] More general theories of how cortical circuits process complex spatial and temporal patterns have been proposed. One such class of models, referred to as *state-dependent networks*, was proposed by my colleagues and me, and later by the Austrian mathematician Wolfgang Maass and his colleagues.[10] To understand this theory it is necessary to understand the concept of the *state* of a cortical circuit.

In physics we can think of the state of a system as the value of the variables that provide the relevant information about the current "configuration" (the state) of that system. If we consider a bunch of billiard balls on a table, its state can be defined by the position and momentum (the mass times the velocity) of each ball. In principle, knowing the state of the billiard balls at some point in time provides all the information necessary to predict not only what will happen next but what happened in the past: knowing the state at time t, the laws of physics allow us to determine the state at time t-1 and t+1. What is the equivalent set of variables that would allow us to define the state of a group of neurons within the brain?

Typically, the state of a network of neurons at a given moment in time is defined by which neurons are firing. I will refer to this as the *active state* because it determines which neurons are actively transmitting information to their partners. But this a highly incomplete description of the state of a neural network, because it is not possible to predict what a circuit will do in the near future solely based on the current active state. There are many other relevant neural properties that influence the future behavior of a circuit. One such property is short-term synaptic plasticity. Clearly what a group of neurons will do next depends not only on which neurons are currently firing but on the effective strength of each synapse at any given moment in time—which depends on what those synapses did in the past. Short-term synaptic plasticity is simply one of many different neural properties that can vary over the time span of hundreds of milliseconds. I will refer to these properties as defining the *hidden state* of a network, hidden because they are concealed from the probing electrodes of neuroscientists.

The active state of a network at a given time t is governed by the input to the neural network and its state (the active and the hidden state) at time t-1. Once again the ripple analogy comes in handy. Consider two raindrops that fall into the pond, the first at t=0 and the second at t=100 ms. The state of the pond at t=101 ms will depend on the interaction between the input (the second raindrop) and the current state (the waves left by the first raindrop). Importantly, the pattern of waves produced by the second raindrop will be different if it falls 100 or 200 ms after the first. The bottom line is that if we were shown a snapshot of the ripples on the pond taken at $t=400$ ms, we could not only tell if one or two raindrops fell in, but if two, the interval between them; the recent past experiences of the pond are stored in its current state. Similarly, the response of a network of neurons is determined by the current input and what just happened—which is represented or encoded in the current state of the network. The "response" of both the pond and a neural network can be said to be *state-dependent*. Indeed, recordings in the auditory and visual cortex demonstrate that the neuronal response to a stimulus is strongly influenced by the preceding stimuli, and how long ago the preceding stimulus occurred.[11] Computer simulations have shown that, in principle, state-dependent networks can discriminate not only simple intervals, but complex spatiotemporal patterns such as spoken words.[12]

POPULATION CLOCKS

Using the ripples on a pond or the changing state of a neural network to tell time does not really answer a key question: what is the code? That is, how does one translate the ripples on the pond or the state of a neural network into units of time? Experimental and theoretical studies suggest that one way the brain encodes time is by determining which neurons are active at a given moment. We have already seen a simple version of this idea in our discussion of timing in the HVC neurons of songbirds—it is possible to determine the amount of time elapsed since the song started by noting which neurons are active, much as it is possible to determine how long ago the first domino was tipped by observing which domino is currently falling. However, this is a very simple chain-like code; the more general idea is that each moment of time is represented by a large subpopulation of active neurons. I will refer to this way of encoding time as a *population clock*. This important concept was first put forth by the neuroscientist Michael Mauk, then at the University of Texas Medical School, Houston. In the nineties Mauk proposed that some forms of timing rely on a dynamically changing population of neurons in the cerebellum—an anatomically distinct part of the brain involved in some types of motor timing.[13] For example, let's assume a stimulus such as an auditory tone presented at $t=0$ triggered a pattern of neural activity in the cerebellum. The idea is that 100 ms later a subpopulation of thousands of neurons might be active, and at time $t=200$ ms another subpopulation of neurons might be active. Even if some neurons are active at both time points, and no single neuron conveys the time, the population clock still allows one to determine if 100 or 200 ms have elapsed.

As an analogy, imagine looking at the windows of a skyscraper at night, and that in each window you can see whether the light in the room is on or off. Now assume that the people in each room have their own quirky schedule, which is repeated every night. In one window the light goes on immediately at sunset, in another an hour after sunset, and in another the light goes on at sunset and then off after an hour and then back on in three hours. If there were 100 windows, we could write down a string of binary digits representing the state of the building at each moment in time: 1 0 1 . . . at sunset, 0 1 1 . . . one hour after sunset, and so forth—each digit representing whether the light in a given window was on or off. Which windows are "on" (1) or "off" (0) at any given moment represents the *state* of the building (the equivalent of the active state of a neural network). We can represent this state as points on a plot in which each axis represents a single window. The problem, of course, is that we'd need a plot with 100 different axes. Figure 6.4 shows how we could represent the state of the building (each point in time) if we had only three dimensions, in which the first, second, and third digit represent the x, y, and z axes of a 3D plot. Although it is not possible to visualize such a plot in 100-dimensional space, the principle is exactly the same. By connecting the dots at each point in time, we can visualize the *trajectory* of the building: how the state of the building changes over time. So even though the building was not designed to be a clock, you can see that as long as it has its own internal dynamics (changing patterns of lights), we could use it to tell time.

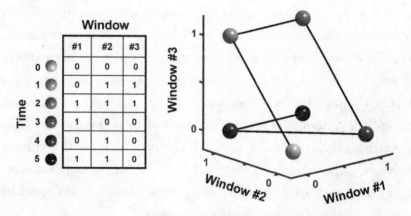

Figure 6.4: Encoding time in the changing states of the windows of a building. The states of the three windows at each point in time (shown in the table on the left) are equivalently represented as a trajectory in 3D space (right).

EVENT-SPECIFIC CLOCKS

We can now see how a changing pattern of active neurons can potentially be used as a timer. But a key insight provided by Michael Mauk's theory is that a circuit composed of a large number of neurons is not just *one* timer, it is many. The advantages of having the same circuit function as a multitude of different timers might not be immediately clear, but this strategy creates a more powerful computational system.

You may have a timer in your kitchen that can be set to time how long to cook a soft-boiled egg, boil pasta, or bake a cake. Another strategy would be to have three different timers, one for each goal—and each timer could have a different alarm sound. Having these three devices on your countertop may seem cumbersome, but this event-specific setup has an important advantage: if you walk into the kitchen and hear an alarm go off, you immediately know what to turn off or take out of the oven. In other words, event-specific timers also serve as a memory of the events that are currently transpiring.

To better understand the value of having multiple timers within a single neural circuit, let's imagine thousands of LED lights wrapped around a Christmas tree, and let's suppose the pattern of illumination of the lights changes in some consistent pattern every time we flip the switch. We can imagine many different types of patterns: a simple chain of blinking lights or, like the windows in our skyscraper, some highly complex time-varying pattern. The advantage of the first, chain-like, pattern is that the code is trivial to read—the first light represents $t=1$, the second $t=2$, ... The disadvantage is that there is only one such pattern—so there is only one timer. In contrast, complex patterns are hard to read, but the same bank of lights could be used to create an enormous number of timers.

Perhaps our bank of Christmas lights has two switches, one controlled by Alice and one by Bob. Perhaps Alice's switch activates the following spatial patterns of lights at 1 sec intervals (where each number represents the position of the light in the chain):

$t=1$	5 10 15 20	
$t=2$	6 12 18 24	
$t=3$	7 14 21 28	

. . .

whereas Bob's switch produces the following sequence of illumination (note that in this example each pattern follows a specific algorithm):

$t=1$	1 2 3 4	
$t=2$	1 4 6 8	
$t=3$	1 6 9 12	

. . .

Now, if you are shown a picture of the Christmas tree and see that lights numbers 8 16 24 32 . . . are on, you not only know that the picture was taken four seconds after the switch was flipped, but that it was Alice who flipped the switch. Our Christmas lights tell time *and* space because they measure how much time has passed and which switch was flipped.

Why would the brain want to have timers that functioned in this manner? Because on the time scale of milliseconds to a few seconds, the brain does not simply need to tell time: it needs to get things done at specific moments in time. One common example of motor timing in animals relates to a form of classical conditioning referred to as eyeblink conditioning: by presenting an auditory tone followed by a puff of air to the cornea 250 ms later, over and over again, humans and other animals will learn to blink in response to the tone. But they do not blink as soon as the tone goes on: the blink is timed to precede the expected air puff. In other words, animals don't only learn to blink, they learn *when* to blink. This is believed to be important because in the situations in which dangerous stimuli might injure the cornea, it is probably not a good idea to keep one's eyes closed for too long.[14] But can animals learn to blink at two different times in response to different tones? This is exactly what Michael Mauk demonstrated. His lab showed that rabbits can learn to time their blinks close to 150 ms after the onset of a low-frequency tone, and 750 ms after a high-frequency tone. Furthermore, after the cerebellum was lesioned, this differential timing effect was gone, suggesting that the neural timers reside within the cerebellar circuits.

As a further example of the importance of having multiple timers, consider a pianist who can play two songs on the piano. The first song requires playing a C note one second into the song, while the second song requires playing an E. A conventional timer would be useful because it tells you when one second has elapsed, but it does not tell you what key to press at one second. In contrast, by using different dynamic spatiotemporal patterns as timers—for example, the different illumination patterns of lights used by Alice and Bob—it is possible to solve not only the problem of telling time, but of determining what to do at each point in time. The brain can implement this strategy by connecting the population of neurons active one second into the first song to the motor neurons responsible for pressing the C key of the piano, and those neurons active one second into the second song to the motor neurons responsible for pressing the E key.

BRAIN DYNAMICS

Neuroscientists have observed many examples of both simple and complex temporal patterns of neural activity that appear to encode time. In a study led by the neuroscientist Joe Paton, at the Champalimaud Center for the Unknown in Lisbon, rats were trained to poke their noses into one of two "ports" depending on whether they were presented with a short or a long auditory interval. On each trial the rats heard two tones separated by intervals ranging from 0.6 to 2.4 seconds. Rats were rewarded for poking their nose into a port on the left if the interval was shorter than 1.5 seconds, and for poking into the right port if the interval was longer than 1.5 seconds. Rats were able to perform the task fairly well; for example, in response to intervals of 1 and 2 seconds they poked into the correct ports around 90 percent of the time. As the rats performed this task, the investigators recorded from dozens of neurons in the striatum—an area of the brain involved in movement and some forms of learning. Many of these neurons fired consistently over many trials at similar points in time during the trial. For example, during the presentation of the 2.4 sec interval some neurons fired early, and others later, and when the activity of the neurons was sorted according to when they fired, a chain-like pattern of activity was observed: A→B→C→D→E—although this description somewhat simplifies the true complexity of the pattern of activity.[15] During trials in which the presented interval was close to the 1.5 second boundary, rats, as expected, were more likely to make errors. Interestingly, it was possible to predict these errors based on the dynamics of the neurons. For example, when the activity patterns were "running fast" (the pattern unfolded more quickly than average), rats were more likely to respond "long," and vice versa. Overall these studies provide compelling evidence that these neurons are contributing to the animals' ability to tell time—although as is generally the case in neuroscience, any single study does not establish that these neurons are actually responsible for timing the interval between the auditory tones.

Similar chain-like patterns of neural activity have been observed in a number of other parts of the brain. For example neurons in the hippocampus of rats can fire at specific moments in time after the animal begins a task, such as running on a wheel, or waiting a preestablished amount of time before making a motor response to receive a reward.[16] Interestingly, in a number of studies, different chains of activity were observed, depending on the details of the task. For example, the same neurons could fire at different points in time depending on the odor that initiated the task, suggesting that these neurons are not simply tracking absolute time but, like the trajectories triggered by Alice and Bob, keeping track of time *and* "remembering" the stimulus that triggered the pattern.

Neuroscientists have also observed much more complex time-varying patterns of neural activity as animals perform temporal tasks. In the case of these complex population clocks, different neurons may start firing at different points in time, each for different amounts of time, and sometimes go back on again later.[17] Such spatiotemporal patterns are reproducible across trials, but might nevertheless appear random to the human eye. It may seem puzzling that there is no apparent rhyme or reason to the spatiotemporal pattern of neural activity in some cases. But this may be the point. One thing we mean by a "random" pattern is that all neurons have more or less the same probability of going on or off at any given moment within that pattern. And we know from the field of information theory that a code in which all the symbols or elements are used with the same probability provides more capacity to store or transmit information. For example, English is not a particularly efficient code, as different letters are used to vastly different extents: if you type in English you probably use the *e* key on your keyboard about 12.5 percent of all key presses, whereas as the *q* key is used only with a frequency of 0.1 percent. Complex, randomish spatiotemporal patterns of neural activity sometimes seem inelegant to neuroscientists, but they might provide the brain with the most efficient way to build a large number of population clocks. Furthermore, it is possible that the brain uses the complex patterns in some areas to drive the generation of the simpler chain-like patterns in other areas.

It is likely that even within the scale of hundreds of milliseconds to a few seconds that the brain employs multiple mechanisms to tell time. Indeed other forms of neural activity have been observed as animals perform temporal tasks. Perhaps the most commonly reported neural signature associated with the passage of time is referred to as a *ramping firing rate*: much like the amount of sand that accumulates over time in an hourglass, the firing rate (the number of spikes within some unit of time) of some neurons increases in a linear fashion over time. Such patterns are typically observed when animals are trained to generate a motor response after a specific delay. But it is not clear whether ramping neurons are actually the timekeepers, or if rather they are reading out the time from other circuits in the brain in order to trigger an appropriately timed motor response.[18]

CHAOS

In our discussion so far we have taken for granted one of the most critical properties of a clock: reproducibility. If spatiotemporal patterns of neural activity within a population of neurons are to be used as a timer, the same pattern must occur time and time again in response to the same context and stimulus. The experimental data from the above studies confirm that this is the case—every time the songbird sings the same neural trajectory is observed (although there is a considerable amount of variability across each trial). A long-standing mystery, however, is exactly how the brain achieves this feat of generating the same pattern time and time again.

Computer models show that neural networks composed of recurrently connected neurons can create a continuously evolving pattern of activity—that is, those patterns that are potentially well suited to encode time. The problem is that such patterns are often not reproducible—indeed such networks often behave chaotically. Mathematically speaking, the term *chaos* is used to describe systems that are highly sensitive to noise and initial conditions (the state of the system at the start of a given trial). The classic example is the weather and the so-called butterfly effect: a tiny event at some point in space and some moment of time, such as a butterfly beating its wings in the Amazon at 12 noon on February 1, can produce a domino effect that changes the weather in New York City a week later. Chaos is often observed in nonlinear physical systems that feed back onto themselves, including the weather or billiard balls. Networks of neurons epitomize both of these conditions. First, neurons are nonlinear—that is, the output of a neuron is not linearly proportional to the input it receives. Second, as mentioned, cortical networks are characterized by a high degree of feedback or *recurrency*—that is, what one neuron does at time $t=1$ will influence what other neurons do at $t=2$, which in turn will influence what the first neuron does at $t=3$.

To understand the problem that chaos poses to using nonlinear dynamical systems to tell time, consider a simple mathematical equation referred to as the *logistic equation* (Figure 6.5). This equation describes the evolution of some value x (bounded between 0 and 1) over progressive time steps. At each step the current value is determined entirely by the value of x at the previous time step. Despite its simplicity, surprisingly complex patterns emerge, and minute vari-

Time Step	Run 1	Run 2
1	0.9900	0.99001
2	0.0386	0.0386
3	0.1448	0.1446
4	0.4829	0.4825
5	0.9739	0.9738
6	0.0993	0.0995
7	0.3488	0.3494
8	0.8859	0.8866
9	0.3943	0.3922
10	0.9314	0.9296
11	0.2492	0.2551
12	0.7296	0.7410
13	0.7694	0.7484
14	0.6920	0.7343
15	0.8313	0.7609
16	0.5471	0.7095
17	0.9664	0.8038
18	0.1268	0.6150

$$\mathbf{x}_{t+1} = 3.9\mathbf{x}_t(1 - \mathbf{x}_t)$$

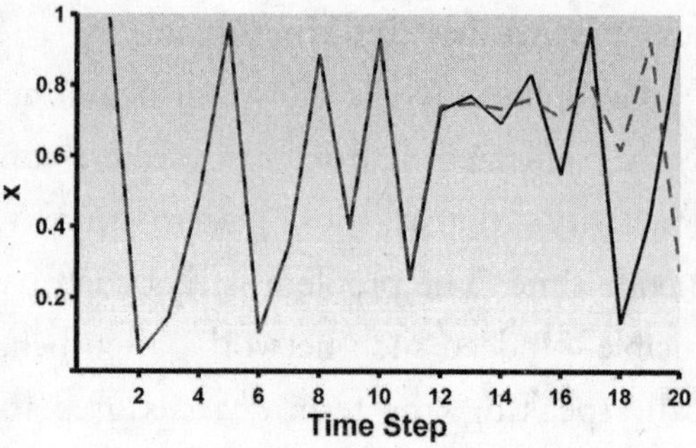

Figure 6.5: Example of an equation that exhibits chaos. In this equation the value of x at each subsequent time step ($t+1$) is determined by the value of x at the current time step (t). Even when starting with two close values of x in Run 1 and Run 2 (0.99 and 0.99001, respectively), the values of x will diverge over time, as shown in the table and graph. The divergence will be imperceptible at first, but after eighteen steps or so the values of x in both runs will be unrelated to each other.

ations in the value of x can produce dramatic differences in future values of x.

Note that we can use the table in Figure 6.5 as a timer of sorts. If you know that that the initial value of x was 0.9900, and I tell you the current value is 0.5471, you know that sixteen time units have elapsed. So in principle we could use a physical system that obeys this logistic equation as a clock. The problem is, however, that this system is extremely sensitive to noise or tiny errors in measurement. For example, if in our second run x started at 0.99001 instead of 0.9900, the value at time step 16 is 0.7095, rather than 0.5471. The state of a chaotic system, the value of x in this example, diverges quickly as a result of tiny perturbations, meaning that in practice the system does not generate the same pattern time and time again. Chaotic systems make really lousy clocks.

Computer simulations of models in which the connectivity between the neurons is randomly determined show that such *random recurrent neural networks* can generate self-perpetuating patterns of activity in which at each time step the network is in a different state. In principle, these spatiotemporal patterns could be used to tell time. The snag is that in the eighties the Israeli physicist and computational neuroscientist Haim Sompolinsky and his colleagues proved that in many cases the patterns of activity that emerge from such randomly connected recurrent networks are chaotic.[19] This posed a serious dilemma for neuroscientists. On one hand the cortex consists of recurrently connected networks that are capable of generating reproducible dynamic patterns of neural activity—otherwise we would not be able to play the same piece on the piano or sign our names in a reproducible fashion. On the other hand, theoretical studies suggested that cortical networks are chaotic.

We do not yet understand how the circuits within our cortex solve the problem of chaos. A number of theories have been put forth to explain how recurrent neural networks can produce complex changing patterns that are not chaotic—that is, that can be triggered over and over again. One such model posits that there are synaptic learning rules in place that essentially allow networks to learn not to be chaotic, rules that allow networks to "burn in" specific patterns or neural trajectories. In theory at least, when the synapses of a computer model of a neural network are appropriately tuned, they can produce complex trajectories that are not chaotic. As illustrated in Figure 6.6, this approach provides a powerful means to generate complex time-varying motor patterns.

The simulation illustrates the activity pattern of ten neurons of a network composed of 800 interconnected neurons, over multiple runs or trials. Each trial begins with a brief input, which sets the initial state of all the neurons in the simulated network. From this initial state onward, a dynamically changing pattern of activity unfolds in time—that is, the network autonomously travels along a trajectory laid out in 800-dimensional space. We can visualize a simplified version of this trajectory in 3D space over multiple trials. Now, to tap into the

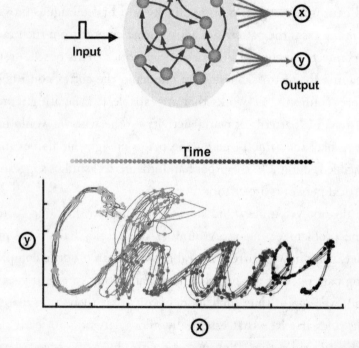

Figure 6.6: A recurrent network that generates a time-varying motor pattern. In this simulation a recurrent neural network is composed of interconnected units representing neurons (schematized in the middle of the top panel). The units in the recurrent network receive a brief input signal, and contact two output units. The activity in these two output units corresponds to the positions of a pen on the X and Y axes of a graph. Training consists of tuning the weights of the connections of the recurrent units onto the output units with a learning rule. After training, in response to a brief input the recurrent network generates a complex pattern of activity that drives the outputs in a manner that writes the word "Chaos." Motor patterns, such as handwritten digits, are inherently temporal, so the network also encodes time. The shaded dots imposed on the lines represent time. The network is not chaotic, as demonstrated by the fact that the motor pattern recovers after perturbing the recurrent network during the upswing of the "h" (ten trials are overlaid). (Modified from Laje and Buonomano, 2013)

computational potential of this network we can connect all 800 of the neurons in our recurrent network to a mere two readout neurons, and stipulate that these are motor neurons that control the movement of a pen on a piece of paper along the x- and y-axes. Although it is rather counterintuitive, as long as the recurrent network driving the two motor neurons is producing a complex (more specifically, "high-dimensional") time-varying pattern, the output neurons can be made to produce almost any pattern (this is achieved by adjusting the strength of the synapses from the recurrent networks onto the output neurons). In the figure this is demonstrated by having the two output neurons "write" the word *chaos*. Importantly, because the recurrent connections were appropriately adjusted, the network is not chaotic. Indeed, it is possible to perturb (or bump) the network in midtrajectory and it will actually return to what it was doing. In essence, there is a memory in the system. The recurrent network has an interesting property of being able to remember what it was doing: even when bumped off its original trajectory it can "return" to and complete the task it was engaged in. Furthermore, note that writing the word *chaos* requires well-timed motor control, and the motion of the pen can be used to tell time. Indeed, the spheres in Figure 6.6 are time markers—knowing the current position of the pen allows us to know how much time has elapsed since the input. The point is that it is possible to "tame" the chaos in recurrent neural networks by tuning the strength of the synaptic connections.

It is productive to pause for a moment and ask: *how exactly does the network write the word "chaos"?* (or, more accurately, where is the information that generates the two-dimensional pattern that the human brain recognizes as the word *chaos*?). This is a profound question that requires one to adjust one's thinking away from more conventional modes of computation and memory. The information that generates the word *chaos* is pretty much everywhere and nowhere. Each synapse and each of the simulated neurons contributes to the pattern, but no single synapse or neuron is actually necessary. The pattern is an *emergent* property: *the whole is larger then the sum of the parts.*

The network described above is only a computer simulation—a simple one with numerous built-in assumptions. Even if this simulation turns out to capture some principles of cortical function, it is much too simple and inflexible to account for the brain's amazing ability to learn to recognize and generate the complex patterns that underlie speech or music. Nevertheless there is increasing experimental support for the notion that many of the computations the brain performs, particularly those that are temporal in nature, rely on the brain's ability to generate complex, time-varying neural trajectories that can be used to produce the spatiotemporal patterns that underlie our ability to reach out and flip the page of a book or play the piano.[20]

The need to tell time permeates almost all tasks the brain must perform, and different tasks have distinct temporal requirements: sometimes it is necessary to discriminate a single half musical note from a quarter note, but in other circumstances it is necessary to tap out a message in Morse code, detect the voice-onset times of the consonants *p* and *b*, or anticipate when the red light will turn to green. To solve this assortment of temporal problems the brain has a suite of interrelated timing mechanisms distributed across its circuits. But interestingly, the clocks within the brain bear little resemblance to the clocks devised *by* the human brain.

The strength of synapses varies over time, the firing rate of neurons ramps up and down, neurons oscillate at specific frequencies, and the activity of networks of neurons changes dynamically over time, because telling time is one thing neurons evolved to do. So asking which neurons or neural circuits within the brain tell time is a bit like asking which of the billion transistors in the CPU of your desktop are responsible for performing binary logic. They all are, that is their raison d'être.

PART II

The Physical and
Mental Nature
of Time

7:00 KEEPING TIME

Time is to clock as mind is to brain.

—DAVA SOBEL

The nucleus of a neuron, and the chromosomes within, are as invisible to the human eye as the moons of Neptune. In the temporal domain, the duration of the beat of a hummingbird's wing is as concealed to our sensory organs as is the drifting of the continents. Much as we build microscopes and telescopes to see objects outside the limited spatial scale of vision, we have developed methods and machines—temporal microscopes and telescopes if you will—to capture time scales far shorter and longer than those the brain can measure. Temporal telescopes have allowed us to establish that humans and great apes split from a common ancestor around 7 million years ago, and predict that in a few billion years the sun will expand into a red giant star and eventually engulf Mercury and Venus. Zooming in, temporal microscopes—high-precision clocks—allow us to split the second into ever decreasing units, milliseconds, microseconds, nanoseconds, picoseconds . . . , each step increasingly beyond the realm of human perception and comprehension. Current-day atomic clocks track time with an accuracy of *attoseconds*—so accurate that, frankly, scientists do not have many other excuses to use the prefix *atto* (10^{-18}). This ability to estimate periods of time on the scale of billions of years, and to subdivide the second into attoseconds, is an outcome of physics, and physics is, in part, an outcome of our desire to tell time.

Astronomy was a seed that bloomed into the science of physics, and astronomy arose from the need to not only locate ourselves in space but in time. Among other things, early astronomy provided a means to track the seasons, establish the length of a year, and determine when to worship the celestial gods. Subsequent advances in our ability to tell time, not coincidentally, overlapped with revolutions within physics. For example, a milestone in the history of clock making occurred in the midst of one of the most transformative periods in the history of physics: the Dutch physicist Christiaan Huygens developed the first high-precision pendulum clock in 1657, fifteen years after Galileo Galilei's death, and when Isaac Newton was a teenager.

Time and physics are inextricably linked. Not only do questions about the nature of time fall within the purview of physics, but it is our understanding of the laws of physics that has allowed scientists to build the stupendously precise clocks that are used to test the laws of physics. In the next chapters we will examine the physics of time and ask if physics and neuroscience are compatible when it comes to their respective takes on the nature of time. But we will start by examining the physics and history of clock time.

OF NEURONS AND NUCLEAR PROLIFERATION

Much as the brain has different mechanisms to tell time prospectively and retrospectively, scientists have devised fundamentally different ways to tell time depending on whether they need to determine the amount of time elapsed from some moment in the past up to the present, or starting from the present to moments in the future. While we generally use conventional clocks to tell time prospectively, we must rely on a different sort of "clock" to tell time retrospectively. Fortunately nature is full of retrospective clocks because the universe is governed by laws that ensure that the changes taking place around us (and within us) obey a prescribed set of rules. Thus the ripples on the pond allow us to peer seconds into the past based on what we see in the present. A forensic pathologist can determine time of death based on the current temperature of a body, and it is possible to determine when two species of animals split from a common ancestor based on the degree of similarity of a given gene. It was the invention of radio-dating, however, that provided one of the most transformative ways to tell time retrospectively. But how does radiodating work? How can atoms "know" how much time has elapsed? To answer this question, we will explore how radiodating and nuclear proliferation helped topple a century-old dogma in neuroscience.

Throughout the twentieth century neuroscientists were convinced that, in contrast to most cells in the body, new neurons were never formed in adult humans. In the nineties, however, there was increasingly convincing evidence in rats and mice that new neurons were born in some parts of the brain—a process referred to as *adult neurogenesis*. But how to determine if this was the case in humans, whose neurons can generally only be studied after death? While we may know the age of a given individual when he or she dies, it's hardly the case that individual neurons come with a born-on-date stamped on them.

Throughout the fifties and early sixties, aboveground atomic bomb testing, mostly by the United States and the Soviet Union, almost doubled the amount of the carbon radioisotope ^{14}C in the atmosphere. These levels peaked in 1963, the time of the Partial Nuclear Test Ban Treaty, and began to decline thereafter. These increases in atmospheric ^{14}C were mirrored in all living organisms, because through photosynthesis plants incorporate carbon into their biochemical pathways, and since the carbon in our bodies comes from plants, atomic bomb testing produced detectable increases of ^{14}C in human DNA. The carbon atoms incorporated into DNA when a cell is "born" (through the division of a precursor cell) can remain in its DNA throughout the life of the cell. So if new neurons are never formed in adults, the neurons of someone born before the atomic bomb testing era should have low levels of ^{14}C today. But if the neurons continue to divide, some of them will have incorporated higher levels, due to the increase in ^{14}C from the late fifties onward. Investigators in Sweden analyzed postmortem brain tissue and found that the great majority of neurons from people born before 1955 had low levels of ^{14}C in their DNA.[1] That is, most neurons were not formed during adulthood. But as predicted from the animal data, there were higher levels of ^{14}C in a subpopulation of neurons in the hippocampus, proving that some adult neurogenesis can occur in humans. This strange intersection between neuroscience and nuclear proliferation and the use of radioactive carbon as a retroactive clock was instrumental in overturning the accepted dogma that new neurons are never formed in adult humans.

Now to return to our question: how can carbon atoms "know" how much time has elapsed? Can a single atom keep track of the passage of time? Understanding the principles of radiodating serves as a reminder that there are many ways to tell time, and that much like the brain, at the technological level we use fundamentally different mechanisms to tell retrospective and prospective time.

Elements are defined by the number of protons in their nucleus. The isotopes of an element refer to the variants that have different numbers of neutrons. Some of these variants are said to be radioactive because they are unstable: over time they will decay to a stable atomic configuration. Unstable can be very stable: ^{14}C, for example, decays with a half-life of 5,730 years. So if we start with 1,000 ^{14}C atoms we can expect to have 500 radioactive carbon atoms 5,730 years later. We can see that the number of remaining radioactive carbon atoms offers a way to tell time retroactively, and thus determine the age of fossils, cave paintings, prehistoric artifacts, ancient manuscripts, and even neurons. But where does the number 5,730 come from? How does a carbon atom "know" when it is due to decay? The answer is that it doesn't. Remarkably, even though carbon radiodating is one of the most reliable ways to tell time retrospectively, a single atom of ^{14}C does not carry any trace of its age.

The principle behind radiodating relies on one of the simplest possible ways to tell time: *chance*. Imagine that there are one thousand people being held captive inside a casino, and that each of them has ten coins. The cruel but statistically inclined captors tell them that their only path to freedom is for a toss of all ten coins to come up with ten heads. If each person were to take an average of one minute for a round of tosses, then based on the number of hostages currently in the room we can estimate how much time has elapsed. Clearly the fewer people in the room, the more time has passed; furthermore, since we can calculate the probability of tossing ten heads, we can estimate the amount of time that has passed. When tossing ten coins, the probability all ten of them will come up heads is 1 in 2^{10} (1/1024). From this number it is possible to calculate that it should take 710 rounds for half the people to gain their freedom: 710 minutes, or 11 hours and 50 minutes.[2] So every twelve hours or so the number of people in the room should drop by a half: if we see 250 people in the room, it is likely that they have been locked up for close to a day.

Radiodating relies on a very similar probabilistic process. Solely for the purpose of intuition, we can think of neutrons in an atom of ^{14}C as continuously trying to escape from the nucleus by breaking through the cloud of electrons engulfing the nucleus (in actuality a neutron releases an electron and an antineutrino, becoming a proton and transforming the carbon atom into nitrogen). There is no ticking or tocking: each atom is essentially sitting there incessantly playing the odds. If we create a single atom of ^{14}C today, there is a 50 percent chance it will decay into a nitrogen atom in 5,730 years. If we come back in 5,730 years and find that it still has not decayed, what is the probability it will decay in the next 5,730 years? It's still 50 percent. Like the ten coins being thrown time and time again, the atom has absolutely no memory or trace of how many millennia it has been gambling.

The ability to tell time using radioactive decay lies in the statistics of the population: the larger the initial number of radioactive atoms, the more accurate our estimate of time will be. A single element does not provide much information about elapsed time, but the population creates a reliable way to tell time. Note that this strategy bears a certain resemblance with the notion of population clocks within the brain, in which the best way to tell time is to look at the subpopulation of active neurons at a given moment in time.

CALENDARS

For the most part our ancestors were much more interested in prospective than retrospective timing. Thus the earliest attempts to track time were primarily calendric in nature. Predicting the phases of the moon, the arrival of winter, and the migrating patterns of potential prey all proved extremely valuable to survival. Diviners, wise men, calendrical priests, and astronomers used the moon, stars, and the natural rhythms of animals and plants—and plenty of superstition—to determine the most propitious day for going to war, planting and harvesting, performing religious ceremonies, marrying, and burying the dead. Determining these times provided power, and with power, of course, came abuse of power. Roman priests in charge of the calendar were apparently not above using their control of the calendar to shorten the duration of the rule of politicians whom they did not favor. According to the author David Ewing Duncan, "the highly politicized college of priests sometimes increased the length of the year to keep consuls and senators they favored in office longer, or decreased the year to shorten rivals' terms."[3]

Not surprisingly, early efforts to anticipate the changing seasons were based on the two most prominent bodies in the sky. Frustratingly however, the sun and the moon follow different schedules. For millennia, the keepers of time struggled with the annoying fact that the solar year is not evenly divisible by the duration of the lunar months: it takes the Earth 365¼ days to loop around the sun (although for most of human history it was thought that the sun was doing the looping), and that the moon takes, on average, 29.53 days to loop around Earth. There are thus 12.4 lunar months in a solar year. In the fifth century BC, the Babylonians came up with a hack: there was to be a 19-year cycle, in which 7 years had 13 months and 12 years had 12 months. The Egyptians and Romans wisely decided to simply ignore lunar cycles for measuring the rhythm of the seasons and duration of a year. But a challenge remained: there is not an integer number of days in a solar year. In other words, the time it takes the earth to spin around its own axis is not an integer of the time it takes to revolve around the sun. If one were to assume the year was composed of 365 days, after a hundred years summer vacation would start 25 days early.

To mediate some sort of agreement between the days and years, Julius Caesar convened mathematicians, philosophers, and astronomers. They invented the leap year. The Julian calendar was composed of 12 months and 365 days, and a leap year with an extra day every four years. In Julius Caesar's honor, one of the twelve months became Julius (July). Albeit a landmark event in human history, the Julian calendar was not perfect. Further tuning was necessary. Over the centuries, calendar time became misaligned with the solar year, because there are not exactly 365¼ days in a year, but closer to 365.2425. To address this drift, in 1582 Pope Gregory decreed that that leap years would be skipped every 100 years and, furthermore, that the skipping of the leap years would be skipped every 400 years. Today we abide by the Gregorian calendar, but to further account for irregularities of the Earth's rotation we have taken to using leap seconds to keep the solar day aligned with standard Universal Coordinate Time. Since 1972 over two-dozen leap seconds have surreptitiously been inserted into our clocks.

THE FIRST CLOCKS

Calendars may tell us what day it is, but not what time of day. To track the time of day humans have been using shadows cast by the sun since at least the fourth millennium BC. Early sundials were simply a vertical stick in the ground with lines demarcating where the sun would cast a shadow throughout the day. On most sundials the day was divided up into twelve intervals, but throughout most of history these "hours" were not necessarily the hours we know today: there were 12 daylight "hours," whether it was a 15-hour summer day or a 9-hour winter day. These sundials told relative time—the hours contracting or dilating across the seasons.[4]

During the Roman Empire sundials were ubiquitous. And so began the relentless shift from free-form time to the rigid discipline of clock time. Grumbling soon followed. In the second century BC the Roman poet Plautus protested:

> *The gods confound man who first found out*
> *How to distinguish hours! Confound him too,*
> *Who in this place set up a sun-dial,*
> *To cut and hack my days so wretchedly*
> *Into small pieces! When I was a boy,*
> *My belly was my sun-dial—one more sure,*
> *Truer, and more exact than any of them.*
> *This dial told me when twas proper time*
> *To go do dinner, when I ought to eat;*
> *But, now-a-days, why even when I have,*
> *I can't fall to, unless the sun gives leave.*
> *The town's so full of these confounded dials, . . .*

There were other ways to tell time, not standard clocks per se, but timers. Clepsydra, or water clocks, marked a fixed duration by having water flow through a small hole to mark how long it took a vessel to fill or empty. And somewhere around the thirteenth century the first true mechanical clocks appeared. Unlike the water of a water clock, mechanical clocks did not freeze in the winter. And unlike sundials they worked at night and on cloudy days. One might reasonably wonder—in the absence of flights to catch, movies to watch, or jobs that required clocking-in—what was the impetus to accurately keep time throughout the day and night? There was something that some people were required to do at regular intervals, rain or shine: pray. Monasteries were highly regimented entities, and as decreed by Pope Sabinianus in the seventh century this regimen included ringing the bell to summon the monks for prayer seven times a day. So clocks were "not merely a means of keeping track of the hours; but of synchronizing the actions of men."[5] Monasteries and churches were the early adopters of mechanical timekeeping technology. Churches grew bell towers, and often it was the monks and priests who were responsible for watching the church clock and ringing the bell to announce the time to the town—provided they did not oversleep:

Brother John, brother John
Are you sleeping? Are you sleeping?
Ring the morning bells! Ring the morning bells!
Ding, dang, dong. Ding, dang, dong.[6]

PENDULUMS

Galileo seems to have been the first human being to notice that the time it took a weight hanging from a string to complete a full swing was almost independent of the amplitude of the swing. But it was only after Galileo's death that his insights were put to use to build clocks. During Galileo's life, however, the properties of the pendulum were used to create one of the first medical devices invented: the *pulsilogium.* It consisted of a weight on a string, which was attached to a horizontal ruler. The ruler allowed the length of the string to be shortened or lengthened. By changing the length of the string, the pulsilogium allowed the doctor to adjust the period of the swing of the pendulum to match the heartbeat of the patient. Thus the length of the string provided a fairly reproducible measure of heart rate.[7]

Christiaan Huygens was the first to use Galileo's insights to build the first high-quality pendulum clocks. A better mathematician than Galileo, he was able to truly comprehend the intricacies of the dynamics of a weight swinging back and forth on a string. Thanks to his mathematical skills and a number of technical innovations, the clock he designed in 1657 represented a quantum leap in timekeeping technology. Before Huygens, the best clocks were off by approximately 15 minutes a day; his clock lost a mere 10 seconds a day.[8] Ten seconds amounts to approximately 0.01 percent of a 24-hour day. This level of accuracy marked a milestone in the history of timekeeping: these were the first clocks designed by the human brain that were better than the clocks within the human brain. As we have seen, the best biological timekeeper, the circadian clock that governs our sleep-wake cycle, has a performance of around 1 percent (15 minutes)—that is, the period of a circadian clock with a mean period of 24 hours will mostly fluctuate between 23 hours and 45 minutes and 24 hours and 15 minutes.[9]

Huygens's pendulum clocks did not, however, solve what was arguably one of the most pressing scientific and technological challenges in the history of civilization: the longitude problem.

At the end of the fifteenth and the beginning of the sixteenth century, European explorers were busy crisscrossing the oceans, discovering new commerce routes, islands, continents, and even circumnavigating the planet. They also spent an inordinate amount of time lost at sea because they could not reliably figure out their longitude—where they were located along the east–west axis. Latitude could be calculated fairly accurately by the angle of the sun at its highest point (local noon). But there was no known way to precisely calculate one's longitude based on the sun, moon, or stars. This had profound economic impact during a period in which Portugal, Spain, France, England, and Italy were competing for the riches of the New World. Crews were decimated by scurvy and starvation while looking for land, captains ran ships aground, and vast treasures sank to the bottom of the ocean. In one such catastrophic accident, in 1707, the British admiral Sir Clowdisley Shovell sailed his fleet onto the Scilly Isles. Four of his five ships, along with approximately two thousand men, were lost. In part as a consequence of this tragedy, Queen Anne of England established the Longitude Act in 1714, and offered a monetary prize of over a million in today's dollars to anyone who invented a method to accurately calculate longitude at sea.

Longitude is about determining one's point in space. So one might ask what it has to do with clocks? Mathematically speaking, space (distance) is the child of time and speed (distance equals time multiplied by speed). Thus, anything that moves at a constant speed can be used to calculate distance, provided one knows for how long it has been moving. Many things have constant speeds, including light, sound, and the rotation of the Earth. Your brain uses the near constancy of the speed of sound to calculate where sounds are coming from. As we have seen, you know someone is to your left or right because the sound of her voice takes approximately 0.6 milliseconds to travel from your left to your right ear. Using the delays it takes any given sound to arrive to your left and right ears allows the brain to figure out if the voice is coming directly from the left, the right, or somewhere in between.

The Earth is rotating at a constant speed—one that results in a full rotation (360 degrees) every 24 hours. Thus there is a direct correspondence between degrees of longitude and time. Knowing how much time has elapsed is equivalent to knowing how much the Earth has turned: if you sit and read this book for one hour (1/24 of a day), the Earth has rotated 15 degrees (360/24). Thus, if you are sitting in the middle of the ocean at local noon, and you know it is 16:00 in Greenwich, then you are "4 hours from Greenwich"—exactly 60 degrees longitude from Greenwich. Problem solved. All one needs is a really good *marine chronometer*.

The greatest minds of the seventeenth and eighteenth centuries could not overlook the longitude problem: Galileo Galilei, Blaise Pascal, Robert Hooke, Christiaan Huygens, Gottfried Leibniz, and Isaac Newton all devoted their attention to it. In the end, however, it was not a great scientist but one of the world's foremost craftsman who ultimately was awarded the Longitude Prize. John Harrison (1693–1776) was a self-educated clockmaker who took obsessive dedication to the extreme.

It was clear to Harrison and others that if clocks were to solve the longitude problem they would have to be mechanically driven by oscillating metallic springs (balance springs). A pendulum was worthless at sea because the movement of a ship would severely alter its swing. Additionally, the temperature fluctuations on land and sea altered the length of the metal bar supporting the weight (the bob) of the pendulum. Indeed, whether of the man-made or the biological variety, changes in temperature pose one of the greatest challenges to high-performance clocks. Thus clockmakers and evolution both faced the challenge of creating clocks that were invariant to changes in temperature. One of John Harrison's early inventions was the creation of the gridiron pendulum, in which the bob was supported by a system of rods of different metals attached in opposing directions. The result was that a temperature-induced increase in the length of one rod was counterbalanced by lengthening in the other direction—keeping the overall pendulum length the same. Harrison's specialty however, was mechanical clocks. And for these he invented the *bimetallic strip*, in which strips of two different metals (each with different temperature expansion coefficients) were joined together. These temperature-sensitive strips could be used to regulate balance springs so they kept a constant period over different temperatures.

As a result of a number of such advances, and his superb crafts-manship, Harrison's legacy rests in building the first marine chronom-eter to meet the accuracy criteria set out by the Board of Longitude. The solution to the longitude problem established a precedent for a technological trend that would be repeated again and again: that the best way to measure space is with a clock.[10]

QUARTZ AND CESIUM

Harrison and other master clockmakers set the stage for over a cen-tury of incremental advances in measuring clock time. But as the nineteenth century turned into the twentieth, there were temporal revolutions underway. With the widespread availability of clocks that lost much less than a second a day, the problem was increasingly how to synchronize all of them. Even the problem of knowing whether two distant clocks were in synch was tricky: how to determine if a clock in Paris and one in Bern both ring at the exactly the same time? The solution lay in two emerging technologies: electricity and radio waves. At the beginning of the twentieth century, there was a focus on *electrocoordination*: using electricity to send signals from a master clock in one location to slave clocks at other locations with negligi-ble delays. Coordinating time was not an esoteric academic matter, but one driven by the railroads, telegraphs, and financial businesses. And as with most practical matters, inventors sought to patent their inventions. Because Switzerland was a hub for time technology, many such patents were submitted to the patent office of Bern. There, from 1902 to 1909, a reportedly diligent patent officer reviewed all sorts of patents, including some relating to the electrocoordination of clocks. In 1905 the patent officer, Albert Einstein, published the paper *On the Electrodynamics of Moving Bodies*, which, in addition to abolish-ing the notion of absolute time, briefly describes a way to synchro-nize distant clocks.[11]

We shall return to the implications of Einstein's work in the next chapter. For now we will focus on the fact that at the beginning of the twentieth century, after centuries of advances, pendulum and mechanical clocks were about to become obsolete—at least as state-of-the-art timekeeping devices. In the 1920s the first quartz-crystal clocks were invented, and two decades after that the first atomic clocks were built.

Accurate clocks are all about their *time base*. The time base of pendulum clock is, well, a pendulum. The time base of a quartz watch is, not surprisingly, a small quartz crystal. When voltage is applied to a quartz crystal it will physically vibrate at a high frequency. The frequency of the vibration depends on many factors, including the type and shape of the crystal, but generally the quartz crystals of digital watches vibrate at 32,768 Hz (a digitally convenient 2^{15}, 1,000,000,000,000,000 in binary notation). These vibrations are counted with a digital circuit to mark off each passing second.

Today even cheap quartz watches can outperform the best mechanical watches. Nevertheless, serious timekeeping is left to atomic clocks. These clocks have a degree of accuracy that would have been inconceivable to Huygens or Harrison. Whereas Huygens's pendulum clock might lose 10 seconds a day, an atomic clock might be off by 10 seconds today if it had been started when the Earth was formed, 4.5 billion years ago.[12]

The time base of an atomic clock is a bit tricky to envision. Atoms, such as cesium, have a *resonance frequency*, the frequency of the electromagnetic radiation that will cause it to "vibrate"—by which we mean that an electron "orbiting" the nucleus will jump to a higher energy level. The cesium 133 isotope resonates when stimulated with microwave radiation at the precise frequency of 9,192,631,770 Hz. In a manner of speaking it is the frequency of this radiation that serves as the time base of atomic clocks, and the cesium atoms play the role of a calibrator that ensures that the frequency is correct. In 1967 an international consortium defined a second as: "*the duration of 9,192,631,770 periods of the radiation corresponding to the transition between the two hyperfine levels of the ground state of the caesium 133 atom.*"[13] The basic unit of time became permanently divorced from the observable dynamics of the planets and placed in the domain of the imperceptible behavior of a single element.

Much as the state-of-the-art clocks of the eighteenth and nineteenth centuries revolutionized marine navigation, atomic clocks revolutionized navigation in the information age. Whether on your smartphone or a missile head, GPS works by determining the distance between at least four satellites and the receiver on Earth. It takes a speed-of-light signal from a satellite 20,000 km (around 12,500 miles) away around 66 milliseconds to reach you. If you move 10 meters (around 33 feet) away from the satellite, the signal will take an extra 33 nanoseconds (0.000000033 seconds). GPS receivers must pick up such tiny differences between the time of transmission and time of arrival. To accomplish this, GPS requires not only sending a bunch of satellites into space, but placing an atomic clock in each one (a wonderful public service provided by the American taxpayers and the US military). By measuring the time differences it takes a signal to arrive from different satellites, a GPS receiver can use a form of triangulation to figure out its latitude, longitude, and altitude.[14] Today's atomic clocks and GPS satellites could not only have told Sir Clowdisley Shovell the position of his ship, but where he was standing on his ship.

SELLING TIME

Our astoundingly accurate clocks were put to work to not only measure the intangible passage of hours and seconds, but to measure what we were doing with our time. With widely available accurate clocks came hourly wages. Toward the end of the nineteenth century a man called Willard Bundy grasped the importance of tracking and recording workers' time for factory managers and invented a method to record the arrival and departure time of factory workers, and so began *clock punching*. The company he founded, International Time Recording Company, merged into Computing Tabulating Recording Company in 1911, which later came to be called International Business Machines.[15]

When Benjamin Franklin wrote *"Time is money,"* he was referring to day wages: an idle day off was money lost in the form of potential earnings. *Time is money* is exponentially truer today. Stock market traders can exploit millisecond advantages for vast monetary gain. And even the simple act of watching TV is a form of temporal-monetary transaction. Viewers relinquish their time doing something they'd rather not—watching commercials—in exchange for "free" entertainment (paid for by subsequent purchases). In the case of some web-based music and video services we can "buy our time back," by paying fees in order to not be subject to the ads.

The sociologist Lewis Mumford argued that *"the clock, not the steam engine, is the key-machine of the modern industrial age."*[16] If the clock was the key machine of the industrial age, it remains a key machine of the information age. Clocks parcel our lives into ever-smaller units of time. Business meetings are timed to the minute; speed daters go on three-minute "dates"; and shaving a fraction of a second off the duration of yellow traffic lights can result in uproar as a result of increased red light violations.[17] But more importantly, the machine that does define the information age, the computer, would not exist without modern clocks. Clocks not only synchronize the actions of men, they also pace the billions of operations computers perform every second.

———

Man's quest to tell time has been, in a sense, too successful. Centuries ago clocks rarely agreed with each other. Today we have come full circle, not because clocks are imprecise, but because they are too precise.

Einstein's theory of general relativity dictates that time as measured by any clock is subject to the strength of gravity. Thus the same atomic clock will tick faster after it is launched into space aboard a GPS satellite (this effect has to be taken into account for GPS to work). Indeed, time as measured by two state-of-the-art optical atomic clocks will drift apart if one is placed on the floor and the other on a table. Which clock then is telling true time? We will see that the question itself is incorrect.

Today we can measure time with more precision than we can measure anything else. Indeed, space has been subsumed by time: a meter is defined as the distance light travels in 1/299,792,458th of a second.[18] But it is not only the resolution and accuracy of modern clocks that is astounding; it is their range. To measure the weight of a grain of salt, a human being, or a truck we require three very different types of balances. In contrast, an atomic clock can be used to measure the nanosecond delays in radio signals from GPS satellites, as well as to time Earth's yearly voyage around the sun—and to add the appropriate leap second when the Earth is running slow (the Earth's rotation can be irregular as a result of geological and climatic events). No device ever conceived, much less created, by man has the accuracy or the range, of modern clocks. But technical feats aside, our ability to measure time has not brought us much closer to understanding the nature of time. Why does time only flow in one direction? Are the past and future fundamentally different from the present? Or does it only seem to be this way because of a deception of the human brain? These are the questions we will address next.

8:00 TIME: WHAT THE HELL IS IT?

> Of all obstacles to a thoroughly penetrating account of
> existence, none looms up more dismayingly than "time."
> Explain time? Not without explaining existence.
> Explain existence? Not without explaining time. To
> uncover the deep and hidden connection between time
> and existence . . . is a task for the future.
>
> —JOHN WHEELER

The human brain has figured out how to build atomic clocks, crack atoms open, travel to the moon and back, transplant organs and genes from one creature to the next, and has even begun to untangle its own inner workings. These impressive feats sometimes lead us to forget that we are merely unusually smart apes.

The brain is a product of the rather haphazard principles of evolutionary "design." For most of the past 70 million years or so, the brains of primates have been shaped by evolution to dexterously use opposable thumbs, recognize objects, identify each other, and develop social skills and bonds that ultimately enhanced survival and reproduction. It seems safe to say that during this process there was little selective pressure to learn how to read, or derive Pythagoras's theorem. That we are able to do these things is a testament to the open-ended computational abilities of the human brain. Yet the brain has a surplus of bugs, limitations, and inherent biases. As a gratuitous example of a task that the brain performs poorly, try mentally adding the following sequence of numbers:

$$1,000 +$$
$$40 +$$
$$1,000 +$$
$$30 +$$
$$1,000 +$$
$$20 +$$
$$1,000 +$$
$$10$$

More often than not people arrive at the answer of 5,000, instead of the correct answer of 4100. Why is the brain so poor at simple numerical calculations when by any measure recognizing a face or reading this sentence is a far more complex computational task? The standard, but partial, answer to this question is that there was little selective pressure to perform numerical calculations. The full answer runs a bit deeper. The building blocks of any computational device— be it the brain or a digital computer—shape which tasks it is well suited (or ill suited) to perform. No human will ever outperform the simplest calculator in long division, because neurons are slow and noisy computational elements. They lack the speed and switch-like properties of the transistors that form our digital computers.[1]

Our poor ability to perform numerical calculations, memorize random strings of words, or rapidly intuit the probability of two coins coming up heads after throwing four coins into the air are a few of the types of tasks the brain is poorly suited to perform. In the face of these facts, we should probably also ask to what extent the brain's inherent limitations and biases constrain the progress of science. How does the brain's architecture shape our ability to answer questions that it did not evolve to address? Among the many things the brain certainly did not evolve to understand was the brain itself. Another is the nature of time.

PRESENTISM AND ETERNALISM REVISITED

Humans have embarked on a quest to measure time with ever-increasing accuracy. As we saw in the previous chapter, it has been an absurdly successful quest, but the success in measuring clock time has not been accompanied by any agreement about exactly *what* we are measuring.

What is time? I don't mean *clock time* or *subjective time*, but rather the word *time* in perhaps its deepest sense: the nature of time. Philosophers and physicists hold many different theories.[2] Some of these theories are mutually exclusive, and some are subtle twists on a theme. But, as we saw in chapter 1, for our purposes *presentism* and *eternalism* capture the two main views.

As a reminder, according to *presentism* only the present is real: all that exists exists in the perpetual present (in my use of the term, presentism does not imply that time is absolute). The past refers to a configuration of the universe that no longer exists, whereas the future represents a yet-to-be-determined configuration. Under *eternalism*, time has been *spatialized* into a full-blown dimension in which the past, present, and future are equally real. The universe becomes a four-dimensional "block" with one temporal and three spatial dimensions—the so-called *block universe*.[3]

Language poses a long-standing impediment to unambiguous conversations about presentism and eternalism. For example, words like "real" and "exist" can have very different meanings depending on whether one is speaking under the umbrella of presentism or of eternalism. In the context of presentism, the statement "*dinosaurs exist*" is false. But under eternalism, one might argue that the statement is true, because dinosaurs do exist at some other moment in time, a moment that is equally real as the moment you consider to be *now*. So rather than try to define presentism and eternalism in terms of words such as *real* and *exist*, we might be better off defining *exist* and *real* according to presentism or eternalism. Under presentism *real* means it *exists now* and only *now*, because the present is the only moment anything can exist in. In contrast, for an eternalist *real* can refer to something that *exists* anywhere/anywhen within the entire block universe, including dinosaurs and your future descendants.

Language assumes an inherently presentist perspective. As in presentism, in verb conjugation the present is a privileged frame of reference. Indeed, the terms presentism and eternalism are related to what some philosophers refer to as *tensed* and *untensed* time respectively. Tensed time is always grounded in the present: the sentence "I went to the gym this morning and yesterday morning" defines past events in relation to the present. The statement is true today, but it won't be true tomorrow (trust me), and it was certainly not true a hundred years ago. In contrast, a dry inventory of events such as "8AM January 1st 2016, at gym; 8AM January 2nd 2016, at gym," is an example of untensed time. If this list is true today it will still be true tomorrow, and in a sense would even be true a hundred years ago. The events now seem to coexist along some continuum, much like the adjacent squares representing "adjacent" days on a calendar. It is as if time has been spatialized.

TIME, WHO NEEDS IT?

There are theories about the nature of time that do not fit neatly into presentism or eternalism. For example, the physicist George Ellis, among others, advocates for a compromise: a four-dimensional block universe that only contains the past. Under this so-called *evolving block universe* theory, the present is the wave front that progressively freezes an undetermined future into an ever-growing and unchangeable past.[4]

Others believe that time is merely an abstraction, a very useful concept to help explain how the universe works, but unlike mass or energy, time would not be a fundamental ingredient of physics. To understand this view a bit better, recall that, in practice, clock time is always measured by change. No matter how accurate or inaccurate, clocks are always quantifying change of some physical phenomenon. The consequence of this fact is that it is always possible to express time as some other nontemporal physical measure. For example, quartz clocks and watches often mark the time with dials that revolve around a circular face, and when the minute hand goes from 12 to 6 we say 30 minutes have passed; but couldn't we just as well say 180 degrees have passed? Or in the case of a pendulum clock with a base frequency of 1 Hz, instead of saying 30 minutes have passed, we could say 1,800 swings have elapsed. Indeed, the standard unit of time (the second) is not actually defined as some pure unit of time, but instead as 9,192,631,770 cycles of the radiation corresponding to the resonant frequency of cesium 133, which is approximately equivalent to how long it takes the earth to spin 1/240 of a degree around its axis. A hundred and twenty-six of these seconds correspond to the amount of time it took downhill skiers to change their position from the top to the bottom of the mountain during the 2014 Winter Olympics. The point is, clock time can be seen as a convention by which we standardize change. Time provides an incredibly useful way to establish equivalent relationships between the rate of change of different physical systems (adherents of this view are sometimes referred to as *relationalists*). As the physicist Ernst Mach put it in the nineteenth century: "It is utterly beyond our power to measure the changes of things by time. Quite the contrary, time is an abstraction, at which we arrive by means of the changes of things."[5]

The notion that time is a measure of change in the state of physical systems was implicit in the previous chapters on how the brain tells time. Just as we could use the ripples produced by a raindrop falling into a pond as a timer, we saw that the brain can use the dynamics of neural networks to establish correlations between internal network states and changes happening in the external world. So the task of tapping your finger every second ultimately comes down to matching changes within your brain to those of a man-made clock. In the end this is essentially all we mean when we say that the brain is telling time.[6]

The hodgepodge terms and theories about time—presentism, eternalism, tensed time, untensed time, the evolving block universe, relationalism, etc.—is a symptom of the fact that there is no consensus as to what time actually is. Nonetheless, to the extent that there is a favored theory in physics and philosophy, it is certainly eternalism. Eternalism, however, is not merely counterintuitive—it mocks one of the most universal features of human experience: that the present is the interface between a past that no longer exists and an open future that is yet to be. Eternalism does not conform to our subjective feeling that time flows, because all moments in time are as real as all locations in space. So physicists and philosophers must have very good reasons to embrace eternalism. Here and in the next chapter we will examine two of these reasons. Here's the preview: (1) according to the laws of physics the *now* is as arbitrary a moment in time as the *here* is a point in space; (2) Einstein's theory of special relativity seems to imply that all moments in time are permanently laid out along the temporal dimension of the block universe.

AGNOSTIC ABOUT THE PRESENT

Our success in deducing the fundamental rules of the universe is perhaps the crowning intellectual achievement of the human species. The laws of physics are so powerful that they have dethroned the gods themselves as the source of the answers to the questions that haunted early humans. What are those points of light embedded in the night sky? Why does the sun rise and set? Eclipses, natural disasters, and the capriciousness of the weather are no longer attributed to the thousands of deities we have worshiped over the millennia.

Newton took the first step. He described the laws that govern the behavior of the objects that inhabit our daily reality—from falling apples to planetary motion. Einstein, through his theories of special and general relativity, expanded upon (and corrected) Newton's laws. Einstein provided us with the tools with which to grasp the cosmic events after the Big Bang, to understand that space and time are not independent, and to explain gravity not as a force per se, but as the curvature of spacetime. Unlike planets and stars, however, subatomic particles seemed to have their own private rule book—one that flouted Einstein's laws. Over the first decades of the twentieth century the field of quantum mechanics decoded these rules. It is within this spooky quantum world that particles exist in superimposed states (seemingly occupying multiple points in space at once) and where entangled particles seem to instantly affect each other, even if they are light-years apart.

But despite their astounding success and transformative impact on our lives, the laws of physics fall embarrassingly short when it comes to explaining what is perhaps the most reproducible observation humans have ever made: the present is special. As explained by the contemporary philosopher Craig Callender: "The equations of physics do not tell us which events are occurring right now—they are like a map without the 'you are here' symbol. The present moment does not exist in them, and therefore neither does the flow of time."[7]

The fundamental laws of physics also have nothing to say about why time seems to be so committed to moving forward. The equations penned by Newton and Einstein, those that describe electricity and magnetism (Maxwell's equations), and the equation that captures the quantum world (Schrödinger's equation) are indifferent as to whether events unfold in the forward or reverse direction.[8] These equations are said to be *time symmetric*, meaning that much in the same way that driving between Los Angeles and San Francisco is an equally realistic proposition as driving from San Francisco to Los Angeles, Newton's laws are open to events unfolding in the forward or reverse direction. Imagine being shown a movie of the wondrous dance of the moon revolving around the Earth as the Earth revolves around the sun. It is possible to use Newton's laws to mathematically describe this dance. That is, we can write down a set of equations that can be used to simulate the motion of these three bodies. Let's say that after having done this, we were later told that the movie upon which we had based our equations was accidentally played to us in reverse. Would we have to throw away all our work? No. Our equations would still be valid: all we would have to do is flip the sign of the variable t to account for the actual forward orbits.[9] Similarly, if we later found out the movie was from a thousand years ago, our equations themselves would once again be entirely valid. Newton's laws are as agnostic to the direction of time as they are to the past, present, and future. And the same is true of the equations of relativity and quantum mechanics. The laws of physics do not assign any special meaning to the direction of time, or to any particular moment in time: the past, present, and future all stand on equal footing.

TIME'S STUBBORN ARROW

You may be thinking: *Sure, I can see how the equations that govern the dynamics of the planets can run in the forward or reverse directions— after all, a movie of the orbits of the planets looks equally plausible when played forward or backward. But surely the laws of physics must forbid those things that I know from experience to be impossible. Balloons do not unpop, dropped glasses do not unbreak, and the ice cubes in my iced tea do not unmelt. The laws of physics presumably ensure that these things cannot happen!* Surprisingly, they don't.

The standard answer to the mystery of time's arrow was devised by the nineteenth-century Austrian physicist Ludwig Boltzmann. His statistical interpretation of the second law of thermodynamics states that the entropy of an isolated system has a relentless tendency to increase with the passage of time. We can think of entropy as corresponding to the degree of disorder. For example, if we throw ten dice into a box and shake it up, the placement of the dice will be disordered or "randomish," so we can say that the box is in a state of high entropy. But if we carefully balanced all ten dice on top of each other to form a column, the system can be said to be in a highly ordered configuration, in a state of very low entropy.

To understand what exactly entropy has to do with the arrow of time, let's first place two hydrogen atoms in the left side of a box. We later return, and ask *what is the most likely arrangement of the two atoms in terms of whether they are located on the left or right side of the box?* There are three possible states (configurations) in which we may find the box: both atoms on the left (LL), both on the right (RR), or one in each half of the box (since the atoms are indistinguishable from each other, the LR and RL states are one and the same). The probability of each of these states, is ¼ LL, ¼ RR, and ½ LR(RL). So the most likely state is the one in which the atoms are evenly split, because there are two ways to have one on each side. If we peek once again into the box after we have seen this evenly split state, the chances that the state of the box will have "reversed" to its initial state is actually quite high: upon our second peek there is a ¼ chance the box will be back in its initial state in which both atoms are on the left side of the box. If this box with two atoms comprised the entire universe, we might say that the universe went back in time: it returned to a state that is indistinguishable from its initial state (at least at this coarse level of analysis in which we don't care about the precise location of the atoms).

But if we instead released 10,000 hydrogen atoms (still a tiny number of atoms) into the left side of the box, and waited for the system to reach a state in which approximately half the atoms were on each side of the box, the probability of all of them going back to the state where they are all on the left is now inconceivably tiny:

much, much smaller than one in a googol (a googol is 10^{100}, larger than the number of particles in the universe). So when we say that it is unlikely that the atoms in the box will return to the original state, we are not talking winning-the-lotto-unlikely, or even winning-the-lotto-every-week-for-a-month unlikely. We're talking wind-blowing-the-winning-ticket-into-your-living-room-every-week-for-a-month unlikely (admittedly I actually have no idea how to go about calculating what that probability would be; the point is that it will not be happening). The statement that it is extraordinarily unlikely that the atoms in the box will ever return to the original state is actually very profound, because it can be read as meaning that atoms in the box will not "travel back" in time, thus providing an arrow of time.

The second law of thermodynamics is not a law in the same sense that the conservation of energy is. Rather it is a statistical assertion that while reversing the state of an isolated system is ridiculously improbable, it is nevertheless legal. So if you knocked a glass on the floor, and later witnessed it piece itself back together and jump back onto the counter top, this would not actually violate the fundamental laws of physics—Newton and Einstein would not need to turn over in their graves. Why not? As the glass dropped, the potential energy from the glass hitting the floor is transformed into kinetic energy in the form of increased motion of air molecules (thus the sound of the glass breaking). Thanks to the law of conservation of energy, the total amount of energy is preserved (no exceptions), and at least in principle, nothing prohibits all those air molecules from eventually taking on the reverse configuration, and applying the same amount of energy towards bumping all the shards of glass back into each other and placing the intact glass back on the table.

So the second law of thermodynamics does not forbid balloons from unpopping, glasses from unbreaking, and ice cubes from unmelting, but it does the next best thing: it virtually ensures that they won't. Thus the so-called *entropic arrow of time* seems to provide a pretty good explanation as to why all the events in the world unfold according to time's arrow. Unfortunately, however, the entropic arrow, in and of itself, is not as much of an arrow as it initially appears to be.

THE DOUBLE-HEADED ARROW

Imagine that our box now has a total of ten hydrogen atoms, and at a particular moment in time (let's call it time t), we observe four atoms on the left and six on the right—we'll represent this state as [4, 6]. We know that the state of maximum entropy is five atoms on each side [5, 5], as there are more ways to arrange ten atoms as two groups of five than in any other distribution. So at the next instant of time, $t+1$, we can guess that we are more likely to observe the [5, 5] state—an increase in entropy—than the [3, 7] state. But instead of looking into the future, let's look back into the past and ask what was the most likely state of our box at the previous instant of time, at $t-1$. Well, by the same logic the answer has to also be [5, 5]. In the absence of any other information about the system, if the most likely state after observing a [4, 6] arrangement is a [5, 5] split, it must be the case that the most likely previous state is also a [5, 5] split. Note that in this case I never said that the system started off with all ten atoms on the same side of the box—indeed, perhaps the box started in the [5, 5] configuration and the [4, 6] state was an inevitable fluctuation.

This comes as a bit of a blow. If we are to use the second law of thermodynamics as an arrow of time, it is disappointing to learn that it predicts that entropy should increase going forward *and* retrodicts that entropy should increase going backward in time. The entropic arrow of time seems to be a double-headed arrow. The thermodynamic explanation as to why time appears to be a one-way street only makes sense because a hidden assumption went along with it. In the first examples above, we started off with all our atoms on the left side of the box—that is, in a state of extremely low entropy. If we start in the state of lowest entropy, then entropy can *only* increase. So the second law of thermodynamics only establishes an arrow of time provided that the universe started in a low-entropy state.

Time itself is often said to have begun with the Big Bang around 14 billion years ago, and in the instants after the Big Bang the universe was indeed in a very-low-entropy state. So the question of the arrow of time now becomes: *how did the universe come to find itself in this initial low-entropy state?* Ludwig Boltzmann was aware of the importance of answering this question, and proposed a clever hypothesis: that the low-entropy state of the universe is the consequence of a transient fluctuation within what was once a higher-entropy universe. If this proposal seems to run against Boltzmann's own law, that is because it sort of does. But as mentioned, the second law of thermodynamics is statistical in nature: decreases in entropy are improbable, not impossible, and given enough time the improbable becomes probable. A related but more modern hypothesis to the low-entropy mystery is the *multiverse* scenario, in which our universe began as a local low-entropy region within a much larger multiverse.[10] Nevertheless, there is no generally accepted theory as to why the universe started in a low-entropy state, and, needless to say, questions relating to the beginning of the universe—and thus to the beginning of time—are not likely to be resolved anytime soon.

The second law of thermodynamics provides a potential explanation for why time marches relentlessly forward—or at least why the universe is undergoing a progressive increase in entropy since the puzzlingly low-entropy state at the time of the Big Bang. But there are other hypotheses of the cause of time's arrow. One is that there are time-irreversible processes ("arrows") embedded within quantum mechanics. I said earlier that all the known laws of physics, including the equation that governs the quantum world (Schrödinger's equation), are time-reversible, and they are. But there is an additional stage to quantum mechanics, one not captured by Schrödinger's equation, that has baffled scientists for nearly a century. If we shoot a single electron towards a photographic plate, Schrödinger's equation gives us the *probability* the electron would be observed at any position at time *t*. But to actually *know* where the electron is, a measurement has to be made, and Schrödinger's equation does not describe what happens during the measurement stage itself. Until a measurement is made— for example, by the electron hitting the photographic plate—the electron is said to be in all possible locations simultaneously. It is only the act of measuring the position of the electron that forces it to be in a definite location—the act of measuring is said to collapse the *wave function* of the electron. But what exactly it is about the measurement process that collapses the wave function (or if it even collapses at all) is not agreed upon. Some physicists believe that the measurement stage of quantum mechanics imposes an arrow of time upon the universe.[11] Within this interpretation of quantum mechanics, once the position of the electron is measured, there is no going back. In fact, once the measurement is made, it is impossible to use Schrödinger's equation to retrodict which slit the electron went through.[12]

Even if it turns out that quantum mechanics imposes an arrow of time upon the universe—something many believe is unlikely— the fact remains that quantum mechanics or any other of the laws of physics do not assign special significance to the *now*.[13] The fundamental equations of physics seem to imply that *now* is to *time* as *here* is to *space*, providing one reason many physicists and philosophers believe that we live in the block universe of eternalism. But to many people, myself included, this is not the most compelling argument in favor of eternalism; rather, as we will see next, Einstein's theory of relativity is probably the best reason to embrace eternalism.

9:00: THE SPATIALIZATION OF TIME IN PHYSICS

> For us believing physicists, the division into past, present and future has merely the meaning of an albeit obstinate illusion.
>
> —ALBERT EINSTEIN[1]

One thing that makes basketball so exciting is that the outcome of the game can come down to a race against the clock. The player taking the last shot must release the ball before the clock ticks down to zero and the buzzer sounds. If the ball leaves the hand before the game clock hits zero the shot counts. It would seem that determining which of these two events took place first would be an entirely objective endeavor: the ball either left the player's hand before the clock ticked down or it didn't. It turns out, however, that this is not the case.

As a thought experiment let's suppose that a referee determines that a game-winning shot taken at one end of the court did indeed leave the hand of a player before an atomic clock at the other end of the court counted down to zero. Using some high-tech equipment the referee later confirms that there was a full nanosecond (a billionth of a second) left on the clock when the ball was released. Now let's also suppose that because this was game seven of the NBA finals, an astronaut was watching the game through a telescope while on an absurdly fast spaceship traveling half the speed of light.[2] Upon learning that the shot was valid, the astronaut finds himself slandering the ref's dear mother, because the astronaut determined that the clock ticked down all the way to zero *before* the ball was released—and thus the basket should not count. These diverging reports of whether the basket counted and which team was the rightful champ have nothing to do with the delays associated with the amount of time it takes information to travel to the spaceship—we are assuming all parties took those delays into account. The two narratives are simply two equally valid realities, one in which the victorious team deserved to win and another in which the referee gave the game away.

How can this be? Is it possible that two events could occur in one order for one observer, and in a different order for another? If so, what would that say about the nature of time? To answer these questions we must delve into Einstein's theory of special relativity.

SPECIAL RELATIVITY

The humble title of Einstein's *On the Electrodynamics of Moving Bodies* offered no clue that the paper would transform the course of science. Einstein derived the theory presented in the paper—the special theory of relativity—starting from two principles. The first was that *the laws of physics are the same for all observers moving at a constant speed*.[3] Einstein borrowed this so-called *principle of relativity* from Galileo, who described it by pointing out that for an observer inside a ship moving smoothly at a constant velocity, it is impossible to know if she is actually moving or not—perhaps you can relate to this if you have ever groggily awoken on a plane to find yourself momentarily confused as to whether you are in flight, taxiing, or stopped on the runway. As a consequence of the principle of relativity we always define velocity in relation to something else. When we say a car is traveling at 100 km/hr we implicitly mean in relation to stationary objects on planet Earth, such as the speed limit sign that says 80 km/hr. But strictly speaking, there is no correct or absolute reference frame. In relation to the police car zooming by in the opposite direction the car's velocity will be well over 100 km/hr; furthermore, it is equally valid to say that the car is at rest and the billboard is traveling at 100 km/hr. So the speed a given object is traveling is relative to the chosen frame of reference. With one exception . . .

The speed of light in empty space is constant and independent of the motion of the body that emitted it. This is Einstein's second principle. At first glance the notion that the speed of light is constant may sound innocuous enough, but together with the principle of relativity, it demolishes the notion of absolute time. To understand the consequences of the constancy of the speed of light let's first agree on the commonsense notion of velocity. If you are on a train traveling 100 km/hr and shoot a bullet in the direction of the train's movement—from a gun that is known to shoot bullets at 300 km/hr—you will observe the bullet traveling away from you at 300 km/hr. If I were witnessing this while standing on the platform of the train station, I would—intuitively enough—measure the bullet's speed to be the train's speed plus the bullet's speed: 400 km/hr.[4] Next let's consider a similar scenario, but now in the context of Einstein's second principle: the constancy of the speed of light. Your train is now traveling at the absurd speed of 100,000 km/s (one third of the speed of light), and rather than a bullet you pointed a laser beam in front of the train. The leading edge of the laser light will be traveling away from you at 300,000 km/s (approximately the speed of light, represented as c). It stands to reason that while you observe the beam traveling at 300,000 km/s, I should, once again, observe its speed to be the speed of the train plus the speed of light: 400,000 km/s ($1.33c$). This, however, would be a severe violation of the principle of the constancy of the speed of light, which insists that everybody will always measure the speed of light as equal to c no matter their own velocity (it would also violate a related outcome of special relativity, that nothing can travel faster than the speed of light). The fact is that both you and I will report that the light beam from your laser is traveling at exactly the same speed.

On an intuitive level this is highly disturbing. After one second you would calculate that the leading edge of the laser beam has traveled 300,000 km ahead of your train. Because I also observe the beam traveling at the same speed, I will calculate that the beam is 300,000 km ahead of the train station *and*, because I know your train is traveling at 100,000 km/s, the train should be located exactly 100,000 km up the tracks. Thus from my frame of reference the distance between the train and the beam should be the difference between both positions: 300,000 − 100,000 = 200,000 km. But you just observed that the light beam is 300,000 km ahead of you! *Something is awry.* Simply put, the price to be paid for the speed of light being absolute is that space and time themselves must not be! Our calculations don't match up because it turns out that we are not experiencing time or space in the same way.

Nineteen five was Einstein's "miracle year": the year he published four seminal papers while still working as a patent clerk in Bern. In his special-relativity paper he derived a set of equations that describe how time dilates (and space contracts) as a function of velocity. Interestingly, the equations are referred to as the Lorentz transforms because they were first described by the Dutch physicist Hendrik Lorentz. But Lorentz did not fully grasp the consequences of his equations, nor did he realize they could be derived from the two principles mentioned above. It is worthwhile taking a quick look at a reduced version of the Lorentz transformation for time,[5] because it is one of the most important equations about time in the history of time. The equation only involves algebra, and it converts the time given by your clock (t^{you}) as you travel in the train to the time on my clock (t^{me}) as I stand on the train platform (assuming that we both started our stopwatches at the instant you zoomed past me). In the equation, v represents the velocity between us, and the constant c is again the speed of light:

$$t^{me} = \frac{t^{you}}{\sqrt{1 - \dfrac{v^2}{c^2}}}$$

Because c is a huge number, at everyday velocities, the term · will be close to zero, and the denominator will be very close to $\sqrt{1}$, that is, 1. Thus, t^{me} will be approximately equal to t^{you}. This is precisely our normal experience: all our clocks tick at the same rate and stay in synch because even when we are moving, we do so at low speeds (relative to the speed of light). But at speeds close to the speed of light, clocks will tick at different rates in relation to each other. Going back to the example where you are on a train traveling at a third of the speed of light, then after one second of travel as measured by your clock ($t^{you}=1$), t^{me} will equal 1.06 seconds. Not a huge difference, but if you were traveling at a speed much closer to the speed of light, say $v=0.999c$, for a year ($t^{you} = 1$ year), t^{me} comes out to over twenty-two years. We say that time has dilated for you: I have aged twenty-two years while you have only aged one.[6]

One of the first experiments to demonstrate time dilation was performed by taking atomic clocks on commercial airline flights and then comparing their time to earthbound atomic clocks. The clocks logged hundreds of hours on eastbound flights (the direction of the flight matters because of the rotation of the Earth). As predicted by special relativity, the traveling clocks fell behind—by tens of billionths of a second—the atomic clocks that stayed home at the US Naval Observatory in Washington.[7]

This and many other experiments have confirmed that time is not absolute. Newton was wrong—clock time does not "flow equably without regard to anything external."

SIMULTANEITY LOST

Whether by the swing of a pendulum, or the amount of the Period protein in a suprachiasmatic neuron, clock time is always measured by change, and change is a local phenomenon. We readily accept that the rate at which some things change can be influenced by their local environment. Which is pretty much why we invented refrigerators—a tomato in the fridge "ages" at a slower rate than its twin left on the counter. Indeed, time, as measured by a pendulum clock or the circadian clock of a fruit fly, can also be altered by ambient temperature. But temperature affects different clocks in different ways—or not at all. For example, the decay times of the radioisotopes discussed in chapter 7 are pretty much the same at close to absolute zero. In contrast, the effect of velocity on the rate of any and all clocks is absolute and nonnegotiable. Any physical process, whether an atomic clock or the human body, will change at a slower or faster rate depending on the speed it is traveling. While this may be disconcerting, there is an even more disturbing consequence to Einstein's theory of special relativity.

Let's return to our train-and-platform thought experiment and again consider the commonsense world in which everything is taking place at low speeds. Consider the example of shooting bullets in opposite directions from within a moving train. Imagine you are in the middle of a 200-meter-long train traveling 100 meters per second (m/s) in relation to me standing on the platform (Figure 9.1). As the tip of your train whizzes by me, you shoot two pistols, the bullets of which also travel at 100 m/s: one bullet is heading towards the window in the front of the train, and the other towards the back window. From your perspective, the bullets are traveling at the same speed and must traverse the same distance, so both bullets will shatter the windows in the front and back of the train simultaneously—exactly 1 second after you pulled the trigger. From my perspective I will see

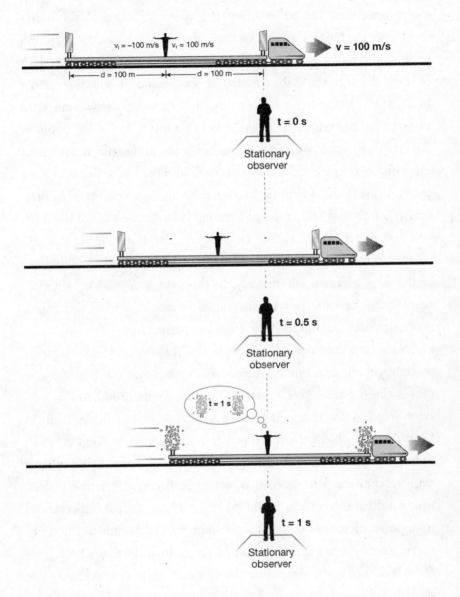

Figure 9.1: Newton's Train. Under Newton's laws, if an observer in the middle of a moving train shoots two bullets in opposite directions (t = 0), the panes in the front and back of the train will break simultaneously for all observers at t = 1 second.

the forward-traveling bullet move at a speed of 200 m/s (the train's speed plus the bullet's speed), and the front window break at 1 second because the bullet had to travel 200 m (half the length of the train plus the distance the train traveled in one second). I will observe the backward bullet moving at 100 m/s (the train's velocity) minus 100 m/s (the bullet's velocity is negative because it is going in the opposite direction). In other words I see the bullet standing still in midair as the window in the back of the train runs into the bullet (it's best if we pretend this is all happening in a vacuum and on a planet with little gravity). This also takes exactly 1 second because the back of the train was 100 meters from where you shot the pistol. As Newton would have expected, both you and I will witness the front and rear windows of the train break simultaneously. In this case we would say simultaneity is absolute: the two events you witness as occurring simultaneously also occur simultaneously from my perspective.

Now let's consider what happens when we perform a similar thought experiment, but at much higher speeds and much longer distances (Figure 9.2). You are now traveling in the middle of a ludicrously long train that you have measured as being 400,000 km in length[8] at two-thirds the speed of light: approximately 200,000 km/s (0.667c). Once again, everything is set up so that when the tip of your train reaches me, you shoot two, yet to be invented, particle pistols, whose bullets also travel at 200,000 km/s. These particle bullets travel in opposite directions towards the windows at both ends of the train. Again, since you are in the middle of the train you will see both windows breaking at the same time: exactly 1 sec on your clock after you shot the pistols, because both bullets had to travel 200,000 km at the speed of 200,000 km/s. And again, I will see the bullet traveling toward the back of the train hanging in midair (because the train's speed minus the bullet's speed is zero), as the back window hurdles towards the bullet at a speed of 200,000 km/s. But at what speed do I observe the forward bullet to be traveling? For both windows to shatter simultaneously from my perspective, the frontward bullet must

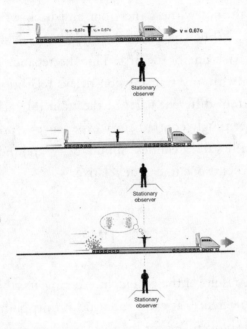

Figure 9.2: Einstein's Train. At high speeds special relativity tells us that different observers will experience space and time differently (making it very tricky to make figures about space and time). The clocks in both the train and platform frames are set to read t = 0 when the front pane of glass reaches the observer on the platform. When the observers on the train and platform are in front of each other, the observer in the train will witness both panes breaking simultaneously, but for the observer on the platform the back pane will have already broken and the front pane will still be intact.

traverse a distance equal to the full length of the train (the initial half length plus the distance the train has traveled) in the same amount of time it takes the back of the train to reach the backward bullet. Since the frontward bullet has to travel double the distance of the backward bullet, it would have to be traveling at speeds well above the speed of light, but special relativity tells us that the speed of the forward-traveling bullet will be around 277,000 km/s (0.92c). So clearly I will not witness the front and back window break at the same time. While you see both windows break simultaneously, I will see the back one break first! This discrepancy has absolutely nothing to do with any transmission delays relating to the time it takes signals from different parts of the train to reach me or you;[9] rather, these seemingly contradictory experiences represent two distinct, but equally valid, realities. Simultaneity, and indeed the order in which two events occur, can be relative.[10]

SPACETIME

Let's digest the results of these thought experiments a bit more. From your frame of reference, at every moment in time both windows are always intact *or* broken. Yet for me there will be a moment in which the back window is broken but the front one is not. This should be deeply disconcerting. How can both panes exist in a broken state for you, but one of them still be intact for me? It is as if we are living in alternate universes.

One resolution to this puzzle is the spatialization of time—that is, the block universe. If we assume that all events that have ever or will ever occur are permanently located at some point in the block universe—as postulated by eternalism—then the relativity of simultaneity becomes no more puzzling than the fact that two objects in space can appear to be aligned or not depending on where you are standing. Two telephone poles along a highway appear aligned if you are standing on the side of the road, but not if you are in the middle of the road—it is a question of perspective. Similarly, both windows can appear to break simultaneously, because they are "aligned" in spacetime from your perspective but not mine. This is why the special theory of relativity provides one of the most compelling arguments for eternalism.[11]

Interestingly, when Einstein first published his special theory of relativity paper he did not argue that time should be thought of as the fourth dimension of a block universe. It was Einstein's professor in Zurich, Hermann Minkowski (who reportedly believed Einstein to be a "lazy dog" during his student days), who first grasped the radical implications of special relativity for the relationship between space and time. In 1908, after building upon his former pupil's work, Minkowski grandiosely announced. "Henceforth space by itself, and time by itself, are doomed to fade away into mere shadows, and only a kind of union of the two will preserve an independent reality."

Minkowski had fused space and time into spacetime. He developed a geometrical reformulation of Einstein's special theory of relativity—one in which there were the standard three spatial dimensions and an additional temporal dimension. Minkowski's insight was that although space and time are relative, an amalgamation of space and time is absolute. If you embark on some zigzaggy voyage in your spaceship while I remain on Earth observing you from afar, upon your return our clocks will disagree about how much time you have been gone and how much distance you have traveled, but we will agree on how much "distance" you have traveled in spacetime. We can simplify Minkowski's four-dimensional universe into a single spatial dimension represented as the horizontal axis of a graph, with the temporal dimension as the vertical axis. Staying at rest consists of movement along the vertical axis—time is passing, but my position in space is the same. Whereas your spaceship voyage is represented by diagonal movement. Based on the change in position along both axes it is possible to calculate the distance traveled in spacetime—a value on which all observers will agree. This distance is related to so-called *proper time* ("local" time): time as measured by the clock in your spaceship.

Special relativity is called *special* because it applies to a simplified universe in which we can ignore the influences of gravity. After publishing his special theory of relativity, Einstein spent ten arduous years developing a more general theory. The result was his masterpiece—the *general theory of relativity*—in which he established an equivalence between gravity and acceleration. Newton's law of gravitation described the relationship between the force of gravity and mass and distance, but he offered few insights into what gravity *really* was. General relativity offered an astonishing answer: gravity was not really a force per se, but the warping of spacetime. General relativity further legitimized Minkowski's marriage of space and time into spacetime. Some would argue that general relativity provides an even more powerful argument than special relativity in favor of eternalism, because some solutions of the equations of general relativity allow for the possibility of time travel—that is, starting from certain assumptions and initial conditions these equations permit jumping backward and forward in time. A detailed discussion of general relativity is outside the scope of this book—not to mention its author's expertise. Fortunately, however, for our purposes special relativity captures the key argument in favor of eternalism and the block universe.

The notion that the past, present, and future are equally real mocks our perception of reality, so if physicists and philosophers favor eternalism over presentism they must have very compelling reasons to do so. We have now seen three of these reasons:

1. The laws of physics provide no evidence that *now* is any more special than *here*, implying that all moments in time are as equally real as all locations in space.
2. Special relativity establishes that two distant events experienced as simultaneous by one observer will not be simultaneous in another observer's frame of reference, and

thus that all moments in time are eternally laid out within the block universe.[12]

3. There are solutions to the equations of general relativity that imply that time travel is possible, and thus that we live in an eternalist universe in which the past and future are in some sense already "out there."

Yet, despite these persuasive arguments in favor of eternalism, we must acknowledge that the laws of physics fail to account for what would seem to be one of the most robust and unequivocal observations human beings have ever made: *that the present is special and that time does flow.*

CAN WE RECONCILE THE PHYSICS AND NEUROSCIENCE OF TIME?

As the epigraph of this chapter suggests, Einstein was an eternalist,[13] but he also seemed to struggle with the apparent specialness of the present. In recounting a discussion with Einstein the philosopher Rudolf Carnap famously elaborated on this point:

Once Einstein said that the problem of the Now worried him seriously. He explained that the experience of the Now means something special for man, something essentially different from the past and the future, but that this important difference does not and cannot occur within physics. That this experience cannot be grasped by science seemed to him a matter of painful but inevitable resignation. I remarked that all that occurs objectively can be described in science; on the one hand the temporal sequence of events is described in physics; and, on the other hand, the peculiarities of man's experiences with respect to time, including his different attitude towards past, present and future, can be described and (in principle) explained in psychology.[14]

As Carnap hinted, many physicists and philosophers believe the only way to reconcile the notion that we live in a universe in which time does not flow and the fact that it certainly seems to, is to relegate our sense of the passage of time to a trick of the mind.

In practice, physicists can generally ignore the dissonance between the physics and neuroscience of time. The equations of special and general relativity account for the experimental data absurdly well—independent of whether the person using them happens to be an eternalist or presentist. Nevertheless, this "block-universe/time-flow" paradox is a profound one. As the mathematical physicist Roger Penrose has put it:

> It seems to me that there are severe discrepancies between what we consciously feel, concerning the flow of time, and what our (marvelously accurate) theories assert about the reality of the physical world. These discrepancies must surely be telling us something deep about the physics that presumably must actually underlie our conscious perceptions. . . .[15]

Similarly, the physicist and author Paul Davies writes:

> In my opinion, the greatest outstanding riddle concerns the glaring mismatch between physical time and subjective, or psychological time. . . . The overwhelming impression of a flowing, moving time, perhaps acquired through a mental "back door," is a very deep mystery. Is it connected with quantum processes in the brain? Does it reflect an objectively real quantity of time "out there" in the world of material objects that we have simply overlooked? Or will the flow of time be proved all to be entirely a mental construct—an illusion or a confusion—after all?[16]

How could something as self-evident as the flow of time be an illusion perpetrated by the brain? One answer to this question goes something like this: Like a reel of film, we can think of the block universe as a series of static frames. Even though a movie contains many different frames—each representing a moment in time—all the frames can be said to coexist within the reel. Much like the frames of a home movie, you are present in many of the frames of the block universe. In each of these frames your mind has memories of the immediately preceding frames. It has been hypothesized that this integrated access to multiple moments in time within a single moment somehow leads to our subjective sense of the passage of time. The lone-wolf physicist Julian Barbour explains this as observed through the motion of a kingfisher (a bird that, as the name implies, is an exquisite fisher):

> When we think we see motion at some instant, the underlying reality is that our brain at that instant contains data corresponding to several different positions of the object perceived to be in motion. My brain contains, at any one instant, several "snapshots" at once. The brain, through the way in which it presents data to consciousness, somehow "plays the movie" for me. . . . I see, coded in the neuronal patterns, six or seven snapshots of the kingfisher just as they occurred in the flight I thought I saw. This brain configuration, with its simultaneous coding of several snapshots, nevertheless belongs to just one. . . .[17]

As mentioned in chapter 1, Barbour, along with a few other physicists, subscribes to a rather extreme version of the spatialization of time. He takes the block universe of eternalism, cuts it up along the temporal axis, and then spreads the slices around a timeless universe that he refers to as *Platonia*. Barbour argues that all possible moments—that is, all the different configurations of matter that comprise all the moments in time—exist as static *nows*.

In the context of the more standard view of eternalism, the physicist Brian Greene presents a similar idea in order to explain how we perceive time to flow despite being stuck within a slice of the block universe:

> Each moment in spacetime—each time slice—is like one of the still frames in a film . . . to the you who is in any such moment, it is the now, it is the moment you experience at that moment. And it always will be. Moreover, within each individual slice, your thoughts and memories are sufficiently rich to yield a sense that time has continuously flowed to that moment. This feeling, this sensation that time is flowing, doesn't require previous moments—previous frames—to be "sequentially illuminated."[18]

There is no doubt that at each moment in time the brain has memories of the preceding moments. As we have seen in chapter 6, the brain is a dynamical system that encodes each event in the context of the preceding events—if this were not the case it would not be possible to understand speech, as the meaning of each word must be interpreted in the context of the preceding words (and sometimes, as we will see in chapter 12, in the context of the subsequent words). Yet even though the brain has access to the preceding frames from within the current frame, I find implausible the notion that the flow of time is an illusion. Indeed, it is far from clear if this *moments-within-a-moment* solution to the block-universe/time-flow paradox is consistent with neuroscience.

IS THE BLOCK UNIVERSE COMPATIBLE WITH NEUROSCIENCE?

The brain is an illusion factory, and most neuroscientists and psychologists would probably agree that our subjective sense of the passage of time is an illusion. So dismissing the flow of time as a trick of the mind is not unreasonable. The problem is, however, that the word *illusion* can mean different things in physics and neuroscience. When a physicist suggests that the flow of time is an illusion, she is suggesting that it exists only in our minds, and that it is not a feature of the external world. When a neuroscientist states that our subjective sense of the passage of time is an illusion, she is suggesting that, like all subjective experiences, it is a mental construct, but one that represents, however unfaithfully, a physical phenomenon that *does* exist in the external world.

The brain is a product of evolution, and evolutionary success is a fairly stringent test of an animal's ability to implicitly grasp and harness the laws of physics (at least a subset of them). The flight of the kingfisher, for example, would be impossible without the nervous system's ability to use Newton's laws: in addition to applying principles of aerodynamics to control flight and the speed of a dive, the kingfisher must extrapolate into the future so that its position will coincide with the projected position of a swimming fish. Furthermore, vision might not convey the true position of a fish, because of the refraction of light that occurs between water and air; and some animals compensate for this optical effect.[19]

The point is that the nervous system is highly attuned to the laws of physics. This holds true not only for motor control—such as a gymnast's ability to pull off a full-twist double layout—but for our subjective mental experiences as well. Our perceptions of color, music, and odors are examples of subjective mental constructs: *qualia*. They are illusions in the sense that they do not exist in the external world, but they are adaptive because each of them is correlated with real physical phenomena: the length of electromagnetic waves, particular patterns of sound waves, and the chemical structure of molecules, respectively. Yet there is nothing intrinsically "blue" about electromagnetic radiation of 470 nm, nor is there anything inherently rotten about sulfur molecules—indeed different animals and people may find the same odor to be repellent, neutral, or attractive.

To grasp the potential adaptive value of our subjective experiences, let's return to the most intimate illusion the brain bestows upon the mind: *body awareness*. As discussed in chapter 4, if someone were to bring a hammer to your finger, you would be immersed in pain; amazingly, although pain is generated within the brain, it is not perceived as occurring within the brain. Somehow, it is projected out into the external world, to the very location where the piece of meat that is your hand happens to be. The brain is so committed to providing an illusion of ownership of the bone, muscle, and nerves that constitute our limbs that it will sometimes persevere in generating the illusion even if the limb has been amputated. So pain is an illusion in the sense that it is a mental construct. Yet, when you feel pain projected out onto your finger, nobody is suggesting that the hammer is an illusion, or that it did not hit your finger. Thus, the illusion of body awareness is not a gratuitous one; there is a very strong correlation between external events (the hammer hitting your finger) and internal subjective experiences (pain). What better way to protect our most important possession than to endow the brain with the ability to *feel* pain: body awareness is the ultimate integration between mind and body, the most sophisticated interface between computer and peripheral device ever built.

Now that we have some insights into the different meanings of the word *illusion*, and the potential link between subjective experience and physical phenomena, let's examine two arguments against the eternalist notion that the flow of time is a mental construct of a physical phenomenon that does not actually exist.

Evolution. If we live in a presentist universe in which time does flow, we can imagine numerous reasons why it might be adaptive to subjectively feel this flow. Much as our conscious perceptions of color or pain are adaptive because they correlate with important events in the external world, perhaps our sense of the flow of time is adaptive because it correlates with events unfolding in the world. Our subjective sense of the passage of time allows us to not only experience a kingfisher's dive, but to anticipate, play back, and rehearse all the external events that play out in time. Perhaps the feeling of the passage of time was even critical to our ability to project ourselves into the far-flung future and engage in mental time travel (chapter 11). But if, in contrast, we live in an eternalist universe in which the flow of time is unreal, how could perceiving it as flowing have been evolutionarily advantageous? Of course, not every biological trait has to provide an evolutionary benefit, but the fact is that most do, particularly those as salient and universal as our sense of the passage of time. The suggestion that our subjective sense of time is a mental illusion in the deepest sense of the word *illusion* seems to imply that it is a gratuitous one, as opposed to a powerful adaptation that better enables the brain to do its main job: predict the future.

Consciousness and Neural Dynamics. The *moments-within-a-moment* solution to the block-universe/time-flow paradox implicitly assumes that it makes sense to discuss consciousness within a frame. While we do not understand how the brain generates consciousness, there are certainly neural signatures that are tightly coupled to conscious states. For example, the most obvious changes that take place as the brain transitions between consciousness (wakefulness) and unconsciousness (slow-wave sleep and anesthesia) are in the temporal patterns of brain activity, most notably in the frequency of brain oscillations, which are a global measure of the synchrony and timing of neural activity. Sleep is characterized by slow brain oscillations, while wakefulness is associated with asynchronous neural activity and fast brain oscillations.[20] Overall, the little we do know about the neuroscience of consciousness tells us that it is a highly dynamic process, and that discussing consciousness in the context of a single frame may be a bit like determining if a cat is alive or not from a single frame of a movie. Is the animal breathing? Is the heart beating? Are the molecules within each of the animal's cells actively engaged in evolution's antidote to the second law of thermodynamics: metabolism? Life is defined by ongoing metabolic changes; if there is no metabolism, an animal cannot be said to be alive. To determine if an animal is alive or not we need to look not only at one frame of the movie, but the preceding and succeeding frames. Life, however, does not represent an argument against eternalism, because there is no need to restrict ourselves to a single frame when determining if an animal is alive or not (we can wait to see multiple frames of the movie to provide a verdict).[21] A related point pertains to Zeno's arrow paradox: can a flying arrow be said to moving if we look at infinitesimally small time slices? In a sense the answer is yes, because we can define an object's instantaneous velocity. But unlike the arrow, conscious beings must be aware of their own "motion" within these instantaneous frames. So the question is whether a slice of the block universe can sustain the phenomenon of consciousness, or does consciousness require some temporal thickness? That is, is consciousness something that can only exist across time slices—something more akin to music than a static image of a movie frame? Steven Pinker seems to be hinting at the challenge of understanding consciousness within static frames in stating: "Matter is extended in space, but consciousness exists in time as surely as it proceeds from 'I think' to 'I am.'"[22]

We will see that consciousness provides neither a continuous nor linear narrative of the events unfolding around us. Rather it seems to be generated in fits and starts, and conscious awareness of external events can take hundreds of milliseconds to develop. So it remains unclear if it makes sense to talk about instantaneous consciousness, and whether the phenomenon of consciousness is compatible with the moments-within-a-moment solution to the block-universe/time-flow paradox.

The laws of physics do not unambiguously state that we live in a 4D block universe. The block universe certainly provides the most consistent *interpretation* of special and general relativity, but it is widely recognized that even within physics there is no universal agreement about the nature of time. There is an ongoing struggle to create a coherent interpretation of the nature of time across all of physics. There are fundamental differences as to the role of time in general relativity and quantum mechanics, which is why time represents a stumbling block in the search for a theory of quantum gravity—the attempt to unify general relativity and quantum mechanics. And there is certainly no experimental evidence that the past, present, and future are all equally real. Indeed there are few explicit experimental predictions that would even distinguish between eternalism and presentism. The most obvious test would be time travel: after all, any discussion of time travel implicitly assumes we live in a block universe.[23] The equations of both special and general relativity allow for time travel, but only under extremely exotic—if not outright impossible—conditions. For example, faster-than-light communication in the case of special relativity,[24] and wormholes stabilized with negative energy in the case of general relativity. So for now, even though the laws of physics seem to be most consistent with eternalism, we have no direct experimental evidence in support of eternalism, much less proof.

So the question is: do the laws of physics (or our interpretation of these laws) need to adapt in order to explain our conscious experience of the flow of time, or does neuroscience need to figure out a way to explain away our subjective sense of the flow of time? Brian Greene eloquently captures this dilemma:

> Is science unable to grasp a fundamental quality of time that the human mind embraces as readily as the lungs take in air, or does

the human mind impose on time a quality of its own making, one that is artificial and that hence does not show up in the laws of physics? If you were to ask me this question during the working day, I'd side with the latter perspective, but by nightfall, when critical thought eases into the ordinary routines of life, it's hard to maintain full resistance to the former viewpoint.[25]

Deciphering whether the flow of time is a fiction created by the mind or something that eludes the current laws of physics is a uniquely complex problem that lies at the interface of physics and neuroscience. And if this mystery were not sufficiently challenging as it is, there is a further wrinkle to consider: the laws of physics and the human brain are not independent of one another. It is not simply that the inner workings of the human brain must obey the laws of physics, but that our interpretation of the laws of physics is filtered by the architecture of the human brain. If we must question whether we can trust our brain's account of something as self-evident as the flow of time, must we not also question the brain's impartiality in interpreting the current laws of physics? As we will see next, the human species seems to have evolved the ability to understand the concept of time by coopting the circuits devoted to understanding space—in other words, the brain itself seems to spatialize time. This raises a fascinating question: do we gravitate towards certain interpretations of the current laws of physics because of the way the brain represents and thinks about time?

10:00 THE SPATIALIZATION OF TIME IN NEUROSCIENCE

One aspect of Einstein's theory does have a counterpart
to the psychology of time, at least as it is expressed in language:
the deep equivalence of time with space.

—STEVEN PINKER

In 1928 Albert Einstein attended a cross-disciplinary conference in Davos. One of the conference participants was the distinguished Swiss psychologist Jean Piaget, who revolutionized the field of developmental psychology by studying how children learn to reason about abstract concepts such as quantity, space, and time. In reference to Piaget's insight that children undergo stereotypical progression in their understanding of numbers, space, and time, Einstein reportedly stated that Piaget's theory was "so simple only a genius could have thought of it."[1]

In his book, *The Child's Conception of Time*, Piaget wrote, "This work was prompted by a number of questions kindly suggested by Albert Einstein more than fifteen years ago when he presided over the first international course of lectures on philosophy and psychology at Davos." One question was "Is our intuitive grasp of time primitive or derived?" In other words, is our conception of time innate or learned?

Apparently Einstein spent time thinking not only about the nature of time, but thinking about how we think about time—a question that is as profound as any other.

CHILDREN AND TIME

Einstein's special theory of relativity was much in vogue in the first decades of the twentieth century, and it influenced the thinking of scientists in a wide range of fields, Piaget included. Regarding the dependence of time on speed, Piaget wondered if there was a parallel between psychology and physics: "The hypothesis that I want to defend is that psychological time depends on the speed or the movements with their speed"[2] (referring to the speed that objects, or the children themselves, were moving).

To glimpse how time is represented in the mind of a child, Piaget asked children to perform a number of simple tasks. One of these tasks involved two toy snails that moved for a few seconds along parallel lines. For example, a blue and a yellow snail might start at the same position and moment in time, and come to a stop at the same moment in time, but the blue snail would travel further because it was moving at a higher speed. Children between the ages of five and six erroneously reported that the snail that traveled the furthest had stopped later.[3]

Studies by Piaget, and many others after him, demonstrate that children come to understand time—or at least respond correctly to questions about the duration of events—only after they understand the concepts of space and speed. For example, when asked questions about toy trains traveling different distances and speeds for different durations, five- to nine-year-olds are more likely to provide correct answers about distance and speed than duration. Even older children often made mistakes about the amount of time objects were moving. In one study 42 percent of children aged eleven to twelve incorrectly concluded that a toy train that traveled further but in a shorter period of time was the train that had traveled for the longer period of time.[4]

One reason children might understand temporal concepts later in development is that the manner in which we measure and quantify time is inordinately complicated. Units of time are expressed through a complex and arbitrary hierarchy: months consist of somewhere between 28 and 31 days; there are 24 hours in a day; 60 minutes in an hour; and 60 seconds in a minute (no metric system here). Furthermore, the same time can be expressed different ways, eight forty-five and quarter-to-nine are the same thing, yet, eight forty-five can refer to a time in the morning or evening. And if that is not confusing enough, we use modular math to tell time—30 minutes after 8:45 is not 8:75.

Since the language of time, and the conventions we use to tell time, seem to have been designed by some villainous organization for the sole purpose of confusing young brains, it is not particularly surprising that children are slow to grasp concepts pertaining to time. Nevertheless, it is interesting that children master questions about speed before time, because we generally define speed in relation to the seemingly more fundamental notions of space and time. Inspired by special relativity, Piaget seemed to believe in the psychological primacy of speed over time. Somewhat murkily he stated: "Relativistic time is therefore simply an extension, to the case of very great velocities and quite particularly to the velocity of light, of a principle that applies at the humblest level in the construction of physical and psychological time, a principle that, as we saw, lies at the very root of the time conceptions of very young children."[5] By which he is suggesting that children intuitively grasp the notion of relative time, and its dependence on speed.

SPACE, TIME, AND LANGUAGE

In chapter 1 I noted that nonhuman animals have a more fundamental "understanding" of space than time. The simple act of placing food in one's mouth or looking behind a tree for hiding prey requires some sort of internal representation of space—a representation more sophisticated than what is needed to cope with the unnavigable, one-dimensional temporal domain. Mice famously learn to run complex labyrinths. Not only can bees reliably navigate between their hive and flowers, but they even communicate the location of specific flower patches to other bees. Animals are also able to extract spatial cues, such as distance, from their senses in a much more direct fashion than they can extract temporal information. For example, the size of the image of a snake on the retina conveys information about whether the snake is far or near—thus allowing for quick decisions about the appropriate course of action. The nervous system of animals evolved sophisticated ways to represent spatial coordinates, such as up and down, left and right, before it developed the ability to explicitly represent the temporal continuum of past, present, and future.

This line of reasoning is consistent with the theory that our ability to grasp the concept of time was coopted from the neural circuits that evolved to navigate, represent, and understand space.[6] As the cognitive psychologist Rafael Núñez writes, "Over the past four decades scholars have converged on the idea that humans conceptualize time primarily in terms of space—a far more tractable domain."[7]

One commonly cited piece of evidence in favor of this theory is that we often use spatial terms to talk about time. Indeed, the linguist George Lakoff and philosopher Mark Johnson have argued that "the experience of time is a natural kind of experience that is understood almost entirely in metaphorical terms (via the spatialization of TIME and the TIME IS A MOVING OBJECT . . . metaphors)."[8] It is actually hard to talk about temporal durations without resorting to spatial adjectives and adverbs. *That was a refreshingly SHORT commercial. We have been studying time for a LONG time.* Similarly, we also borrow spatial terms to talk about the past and future: *I'm looking FORWARD to your reply*; *in HINDSIGHT that was a terrible idea*; *Christmas is CLOSE to New Year's.*

In English, when we spatialize time we place the past *behind* us and the future *in front*. But while all languages use spatial metaphors, time is not always spatialized in the same way. In Aymara, a language spoken in the highlands of western Bolivia and northern Chile, the word for past, *nayra*, also means *eyes* or *sight*, while a word for future, *qhipa*, also means *back* or *behind*, suggesting a fundamentally different perspective in how the Aymara use space to conceptualize time. Rafael Núñez confirmed the uniqueness of the Aymara's spatial-temporal perspective by studying their gestures during speech. Videos of native Aymara speakers show they often pointed forward when speaking about the "old times," and gestured behind them when referring to the future.[9] This perspective reversal is not as odd as it may initially seem. After all, just as we know what happened in the past, we know what is located in front of us because we can see it; it is the future, and what stands behind us, that is unknown.

WEDNESDAY

Even if we do live in a frozen spacetime block in which time does not pass or flow, subjectively time certainly seems to flow. And language reflects this fact, again, by borrowing from the spatial domain. *Time is PASSING us by. The end of the world is APPROACHING. The day FLEW by.* But who or what is doing the passing, approaching, or flying? Am I moving through time, or am I standing still while the river of time flows by me? Linguistically speaking the answer is both.

Perhaps you have encountered the dilemma of being told that *next Wednesday's meeting has been moved forward two days*. So do you show up for the meeting on Monday or Friday? "Forward" is generally taken to be in the direction of movement. Thus if you are moving through a static timeline, and the timeline is moved forward, the target day will put it farther away, on Friday. But if you are standing still, and we conceptualize time itself to be flowing by you, putting the meeting forward will place it closer to you, on Monday. The first interpretation (Friday) is described as an ego-moving perspective, and the second (Monday) as a time-moving perspective (Figure 10.1).

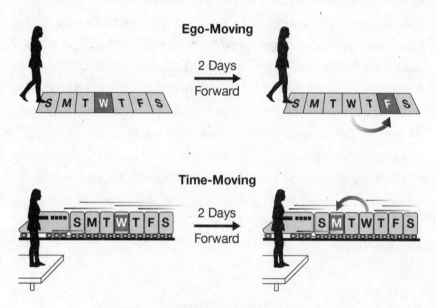

Figure 10.1: Ego-Moving and Time-Moving Perspectives.

This ambiguity is the linguistic equivalent of Galilean relativity: motion must be defined in relation to something. As we have seen, a statement such as *the speed between you and a lion is 10 km/h*, leaves it open as to who is doing moving; indeed, in empty space it does not really make sense to try to determine who is moving toward whom— it's all relative. Nevertheless, in practice it is really useful to know if you or the lion is actively moving; so we might clarify by saying *the lion is running towards you at 10 km/h*. It is implicit that its speed is in relation to our standard frame of reference, the ground. When it comes to statements about moving in time, however, there is no standard reference frame. Studies show that when people are asked what day the Wednesday meeting is on if it was moved forward two days, it's close to a 50-50 split: approximately 50 percent of people assume the meeting is on Monday, the rest think Friday. Fascinatingly, these perspectives are not set in stone. It turns out that the answer depends on people's recent physical movement. For example, when people who were at an airport to pick someone up were asked the Wednesday question, 51 percent answered Friday, but 76 percent of the people who had just arrived answered Friday. The interpretation is that the travelers were in an ego-moving state of mind because they just experienced significant movement through space. Other studies show that physical movement is not actually necessary; simply priming people to think about moving though space can influence the proportion of Monday or Friday choices.[10]

Linguistically, the relationship between space and time is an asymmetric one. Spatial metaphors are often used to talk about time, but temporal metaphors are rarely used to describe space (although occasionally we do use time units to describe spatial distances: *I live ten minutes from here*). This asymmetry has been put forth as evidence that our ability to conceptualize time is built upon our understanding of space. Linguistic arguments alone, however, cannot fully justify such a conclusion. Perhaps we borrow spatial terms to talk about time for more general reasons: space may be a universal source of metaphors because it offers a more natural and richer domain. Indeed, we use spatial metaphors to describe pretty much everything.[11] *We have become very CLOSE since he DISTANCED himself from his brother. Cats have very HIGH standards. I can't tell if you're feeling UP or DOWN today.* It turns out, however, that the link between space and time runs much deeper than language alone. Independent of metaphors or language, space influences how we actually perceive time.

KAPPA

Imagine two lights that are a few feet apart from each other. Each light is briefly flashed on and off, and the interval between the flashes is 8 seconds. You are then asked to reproduce this interval by holding down a button for the estimated duration. The question is: *will the distance between the two lights influence your perception of time* (more specifically, your attempt to reproduce the perceived interval)? One of the first studies to ask this question revealed that when the lights were 8, 16, and 32 inches apart (always flashed 8 seconds apart), the mean time estimates between the flashes was 6.5, 7.15, and 8.05 seconds, respectively.[12] So the answer is *yes*, space (the distance between the lights) does influence our perception of time. This so-called *kappa effect* has been demonstrated many times and in many ways. For example, another study flashed three dots on a computer screen, one to the left, one in the middle, and one to the right. The dots were uniformly spread out in time: the first dot appeared at time $t_1=0$, the second dot at $t_2=0.5$ sec, and the third at $t_3=1$ sec. The subjects were asked whether the first interval $(t_2 - t_1)$ is longer than the second interval $(t_3 - t_2)$. Although both intervals were the same, people's responses were strongly influenced by the distance between the dots: people were much more likely to judge the first interval as longer if the distance between the first (left) and second (middle) dot was larger than the distance between the second (middle) and third (right).[13]

The kappa effect establishes that the distance between two events has a profound effect on people's judgments of the amount of time between them. This relationship between space and time within our brain is further backed by the converse phenomenon. Increasing the time delay between two flashes that are always the same distance from each other causes people to progressively increase their estimates of the spatial distance between them (this illusion is called the *tau effect*).

Although the existence of both the kappa and tau effects seems to indicate a symmetric relationship between space and time, other experiments suggest an asymmetry. Studies performed by the cognitive psychologist Lera Boroditsky demonstrate that sometimes distance influences duration judgments more than durations influence distance judgments. Boroditsky and her colleague Daniel Casasanto asked students at MIT to observe lines that were slowly growing in length on a computer screen. The lines grew to different lengths over periods of time ranging from 1 to 5 seconds. After observing each line, participants were asked to reproduce the total amount of time the line was present *or* the length the line grew to. Results again showed that for the same duration, people's duration estimate was strongly influenced by the distance the line grew. If a line that was present for 3 seconds grew a lot, people correctly judged its duration to be 3 seconds, but if it grew only a little, the duration estimates were closer to 2.7 seconds.[14] In contrast, how long the line was present had little effect on the subjects' estimates of length. Tongue-in-cheek, Boroditsky remarked: "Piaget concluded that children could not reliably distinguish the spatial and temporal components of events until about age nine. Like many contemporary results in cognitive science, our findings suggest that Piaget was right about the phenomenon he observed, but wrong about the age at which children resolve their confusion: apparently MIT undergraduates cannot reliably distinguish the spatial and temporal components of their experience, either."

CLOCK OR MEMORY?

The odometer and clock on the dashboard of your car are incommunicado. Whether you drive 100 km in an hour, or, while embedded in Los Angeles traffic, a mere 10 km, your car's clock will tell you 60 minutes have passed (we are, of course, ignoring the insignificant effects of special relativity). In sharp contrast, the kappa effect seems to suggest that the clock within the brain that is responsible for timing on the scale of seconds is somehow influenced by the brain's odometer—more specifically the neural circuits responsible for estimating distances. But it is not necessarily the case that there is some clock within the brain that gets sped up during the kappa effect; as we have discussed in previous chapters, temporal illusions can also arise as a result of memory distortions.

The experimental paradigms used to study the perception of time always require participants to judge a specific temporal interval *and* to remember that interval in order to compare it to others. So if you were using some sort of neural stopwatch in your brain to solve the first task described above (where two lights were flashed 8 seconds apart at different distances), you would use it to time 8 seconds, proceed to temporarily store that number in memory, and then use the stopwatch again to reproduce the duration stored in memory. Thus the kappa effect could arise not because distance alters the clock speed per se, but because distance alters either the storage or memory of the perceived temporal duration.

It has been proposed that the brain has a multipurpose system responsible for processing information about magnitude[15]—specifically, that there are circuits within the parietal cortex devoted to processing information about quantity, regardless of whether the quantity is spatial, temporal, or numerical. So it is possible that the temporal distortions imposed by space are a result of interactions in how these circuits store information about magnitude. For example, storing a small temporal magnitude and a large spatial magnitude might bias the temporal magnitude upward. We can think of this as a type of a *regression to the mean* effect—having two quantities stored in memory results in both taking on some of the features of the other. Consistent with the theory of a shared magnitude system within the brain, temporal judgments are not only affected by distance: the brightness or size of a stimulus also influences how long a stimulus seems to last. Indeed, as mentioned earlier, studies even suggest that if a low number (say 1) or a high number (9) is flashed on a screen for the same amount of time, people tend to judge the high number as lasting a bit longer.[16]

Additional evidence for the existence of a brain area that represents both spatial and temporal magnitudes comes from the notion of a mental timeline—the mental equivalent of representing time along a dimension of a Cartesian plot (the time axis of a graph). Those of us who have been given the privilege of a formal education often conceptualize time along a line with the short temporal intervals (or the past) to the left, and longer intervals (or the future) to the right. This mental timeline and its relationship to space can be revealed in many ways, including the cryptically named *STEARC (Spatial-TEmporal Association of Response Codes) effect*. Imagine having to listen to a sequence of many tones of different durations, and that after each tone you have to report whether it was shorter or longer than some reference duration, by pressing one of two buttons. It turns out that performance on this temporal task depends on the spatial location of the buttons! People are quicker and better if they have to report the short duration with their left index finger and the long duration with their right index finger compared to the opposite arrangement with the "short" button to the right and the "long" button to the left. In other words, reporting short durations is more natural with your left hand, than with your right hand—as if there were a mental timeline laid out from left to right within your neural circuits.[17] Further evidence for a mental timeline comes again from Lera Boroditsky's lab. People who have suffered a stroke in the right inferior parietal cortex often exhibit spatial hemineglect: they are not fully aware of objects to their left. For example, patients with hemineglect might not eat the food on the left side of their plate, or even fail to groom the left side of their face. Boroditsky and colleagues have provided evidence that hemineglect patients also have a deficit placing information about the past and future along a mental timeline—resulting in impairment in their ability to remember the temporal context of events.[18]

There is also evidence that space and time intermingle at the more basic level of individual neurons.[19] For example, as mentioned earlier, neuroscientists have been recording the activity of neurons that represent space for decades; specifically, from place cells within the hippocampus that fire preferentially when an animal is at a specific location within a room. More recent evidence suggests that a small percentage of cells in the hippocampus can encode the distance a rat has traveled on a treadmill; for example, a "distance" neuron might fire after the rat has traveled five meters—more or less independently of how long it has been running (or equivalently, independently of the speed of the treadmill). Other cells seem to encode the amount of time a rat has been running on the treadmill, perhaps firing after the rat has been running for twenty seconds—again, more or less independent of the distance covered. The vast majority of cells, however, behave in a more complicated fashion: their firing pattern is determined by some complex amalgam of position, distance covered, elapsed time, and speed.

Overall we do not yet understand how neurons in the hippocampus—or anywhere else in the brain—measure, represent, or store spatial and temporal magnitudes. Yet, based on linguistic, psychophysical, and neurophysiological evidence, it is clear that space and time are indeed intertwined within our neural circuits.

RELATIVITY IN PHYSICS AND NEUROSCIENCE

Over the last few chapters we have seen that the relationship between space and time holds intriguing parallels in physics and neuroscience. To recap and expand upon these parallels:

Time Is Relative. Einstein taught us that while the speed of light is absolute, time and space are relative: at high speeds clocks slow down. Einstein also alluded to the relativity of subjective time when he supposedly said, "An hour sitting with a pretty girl on a park bench passes like a minute, but a minute sitting on a hot stove seems like an hour."[20] As we discussed in chapter 4, our subjective sense of elapsed time *is* relative, but it depends on a multitude of factors, including context, emotional state, attention, stimulus features (distance and speed, for example), and whether subjects are under the influence of psychoactive compounds.

Space and Time Are Not Independent. Special relativity imposes a trade-off between space and time: traveling at very high speeds through space brings time to a crawl, while standing still is the "quickest" way to travel along the time axis. Subjectively speaking, space and time are also interdependent. The kappa effect, for example, establishes that when two events separated by the same temporal interval occur at larger distances from each other (reflecting higher speeds), people tend to judge those temporal intervals as longer.[21]

Relativity of Simultaneity. One of the most astounding consequences of special relativity is that two simultaneous events from my perspective are not simultaneous from the perspective of someone in motion relative to me—that is, simultaneity is relative. Although we have not discussed this yet, simultaneity is also relative from a subjective perspective. For example, because of the millionfold difference in the speed of light and sound, visual and auditory signals from the same event arrive at our sensory organs with different delays. Yet while sitting in the cheap seats at the symphony you perceive the sight and sound of the cymbals clashing as simultaneous, even though there can be a delay of close to 100 milliseconds in the arrival of the sound. We will see in chapter 12 that in order to create a convenient narrative of the events unfolding in the world around us, the brain takes the liberty to fudge our perception of what we perceive to be simultaneous.

Piaget, for one, was captivated by such parallels. He seems to have believed in a deeper link between a child's inherently relative notion of time and the relativistic time of Einstein.[22] But any apparent parallel between special relativity and our perception of time is simply that. The interdependence of space and time in physics reveals something profound about the universe, but tells us nothing about the psychology of time. The fact that distance can influence our temporal judgments reveals nothing about the true physical nature of space and time, but it does unveil something profound about the brain's architecture.[23] But what exactly? There are, no doubt, multiple answers to this question. One is that from our first to our last breath the brain records the statistics of what we see, hear, and experience, and uses any patterns it finds to make sense of the world around us. Consider the image in Figure 10.2:

Figure 10.2: Concave-Convex Illusion. We see the middle circle with a dark lower edge as convex (popping out of the page) and the circles with dark upper edges as concave because the brain assumes light comes from above.

Presumably, of the three circles, the middle one provides the illusion of being convex—as if it were rising out of the page—while the circles on either side of it give the impression of being concave—resembling a hole dug into the page (if you turn the page upside down you will see that the circles only differ in their orientation, and that the middle circle now appears to be concave). This illusion is a consequence of the fact that from the day you were born, your visual system has been sampling the statistics of the world: light generally comes from above, so a bump on the wall will cast a shadow on its lower half, whereas a hole will produce a shadow along its upper lip.

Just as your brain uses prior information about sources of light to infer three-dimensional shape, your brain uses its past experiences to make inferences about time and space. We all have a vast database of observations about objects and animals moving through space and time—generally at a fairly restricted range of speeds. Thus, we know that distance and time are correlated: a child watching a drop of rain slide down a window can see that the more time that elapses, the larger the distance the drop traverses. The clocks within the brain are far from perfect and, in the absence of a perfect clock, we use prior experience to guide our judgments. Indeed, the degree to which irrelevant information, such as distance between two flashes of light, influences temporal decisions is relatively small, more or less within the same range as the accuracy of the brain's clocks. What this all means is that if you have an inaccurate stopwatch within your brain, and you need to know for how long two toy trains were traveling, it makes sense to take into account in your estimate the distance each train traveled.[24]

I suspect space and time are entwined within our neural circuits for at least two reasons. First, like evolution itself, the brain has a very opportunistic modus operandi: it's always borrowing and recycling existing features. It is likely that our ability to understand the concept of time was achieved in part by coopting circuits that evolved to navigate and conceptualize space. Secondly, the brain is an expert at scavenging information from patterns in the external world, and since spatial and temporal intervals are strongly correlated, the brain uses distances to optimize its estimates of the passage of time.

Under eternalism, time is spatialized: all moments in time are laid out and frozen within the block universe, leaving us with the conclusion that the flow of time is an illusion created by the mind. But could the illusion be the other way around? Could the architecture of the brain bias our interpretations of the laws of physics?

The physicist Lee Smolin suggests that the progressive spatialization of time in physics has biased our conception of the nature of time: "The ability to freeze time . . . has been a great aid to science because we don't have to watch motion unfold in real time. . . . But beyond its usefulness this invention has profound philosophical consequences, because it supports the argument that time is an illusion. The method of freezing time has worked so well that most physicists are unaware that a trick has been played on their understanding of nature."[25]

Now that we know that the brain itself spatializes time, it is also worth asking if the acceptance of eternalism has benefited from the fact that it resonates with the architecture of the organ responsible for choosing between eternalism and presentism.[26] In other words, since we developed theories and mathematical methods to represent time as a space-like dimension, maybe we are more comfortable with eternalism than presentism because of how the brain conceptualizes time. It is difficult to answer the question, but as we are about to see, the human mind does indeed inhabit an eternalist universe: mentally speaking, not only do the past and future exist, but they are valid travel destinations. Indeed our species is defined in part by our incessant tendency to mentally hop back and forth between the past, present, and future.

11:00 MENTAL TIME TRAVEL

> To be immortal is commonplace; except for man, all creatures
> are immortal, for they are ignorant of death.
>
> —JORGE LUIS BORGES

On March 11, 2011, a 9.0 magnitude earthquake triggered a massive
tsunami that hit the northeast coast of Japan. Approximately 15,000
people were killed, and hundreds of thousands were left homeless.
During the subsequent cleanup operations there were numerous
reports of "tsunami stones": large stones found in some of the devas-
tated areas that had been inscribed centuries ago with warnings like
"Do not build your homes below this point!"[1] These warnings were
heeded, or not, on a case-by-case and town-by-town basis. But the
stones' inscribers were clearly thinking about the distant future, imag-
ining that one day, people like themselves would be faced with the
quandary of deciding where to build their homes. The inscribers were
reaching out into the future and offering advice based on their own
tragic experiences.

The psychologists Thomas Suddendorf and Michael Corballis
have referred to our capacity to mentally project ourselves into the
future as *mental time travel*.[2] And as noted in chapter 2, our ability
to craft a stone into a tool, plant a seed to ensure food in the future,

build a hut, work for a salary, or save for retirement are all contingent on our capacity to envision different futures, and understand that by acting in the present we can sculpt the future. The ability to mentally time travel, to quote again from the psychologist Endel Tulving, "brought with it a radical shift in humans' relation to nature. Instead of using their wits to adjust to the vagaries of nature, including the uncertainties of availability of food, shelter, and protection from predators, humans began to anticipate these problems and take steps to mitigate their unpredictability."[3]

Many psychologists, including Endel Tulving and Thomas Suddendorf, believe that mental time travel is a uniquely human cognitive ability, and indeed that mental time travel is a key ingredient to being human.

REVISITING AND PREVISTING

I have a childhood memory of walking along the edge of the lake in Roger Williams Park. It was winter and parts of the lake had frozen over. Stupidly, as I tested the ice it broke, and I fell into the freezing water. My ability to recall and mentally relive this event relies on two different types of memories: semantic and episodic. The distinction between these two flavors of human memory is sometimes expressed as the difference between knowing and remembering. Semantic memory refers to knowledge, such as the name of the park, and that the park is located in the city of Providence, which in turn is located in Rhode Island. Semantic memory also encompasses knowledge of the even more fundamental facts needed to make sense of this story, such as that water can turn into ice, and that icy water is cold. Episodic memory refers to my ability to mentally re-experience the episode, see the ice in my mind's eye, invoke the emotional content of being cold, and recall that the shallow water made it fairly easy to pull myself back onto the edge.

An often-overlooked distinction between semantic and episodic memory is the absence or presence of a time stamp. You know that water can turn into ice, but I'm going to go out on a limb and bet that you have no idea when you learned this important tidbit of information. You may know the capital of Nepal, but do you know when you learned that it is Katmandu? Our semantic memory stores knowledge about the world, but it does not store the date any particular piece of information was acquired, or even the order in which it was acquired. What did you learn first? *That pickles are cucumbers* or *that raisins are grapes*. In contrast, like the date associated with every file on your computer, episodic memories generally have some sort of time stamp—not necessarily the exact date, but the approximate year, your approximate age, or simply whether the episode occurred before or after another memorable event in your life: if you remember your first kiss and the most embarrassing thing that ever happened to you, you probably know which came first (hopefully those were two different episodes).

Episodic memory and our ability to mentally project ourselves forward in time is heavily dependent on our semantic memory. It would be difficult to mentally time-travel to a planned tropical beach vacation without a working knowledge of sand, sun, oceans, and piña coladas. Consistent with the notion that semantic memories may serve as the infrastructure to lay down episodic memories, developmental studies suggest that semantic memory emerges before episodic memory in children. For example, when four-year-olds are taught the names of new colors, such as chartreuse and taupe, they quickly learn to apply that knowledge when asked to pick up the appropriately colored item. But when asked *when* they learned those color names, they often claimed to have always known the colors they learned just a few minutes ago.[4]

People with so-called *anterograde amnesia* generally lose the ability to store new semantic *and* episodic memories—although they can still learn motor tasks such as learning to ride a bike, and other types of so-called *procedural* or *implicit* memories. Previously stored semantic memories (for example, the names of their family members or the capital of France) are largely intact, but some amnesic patients also have an impoverished ability to recall old episodes of their lives (those that happened before the onset of amnesia).[5]

It is not surprising that someone with amnesia will struggle to describe what he did yesterday—that's pretty much the definition of amnesia. But do people with amnesia struggle to plan ahead or describe what they may be doing tomorrow? The answer to this question seems to be *yes*. Research over the past two decades has progressively emphasized that some amnesic patients struggle to project themselves into both the past and future. One such patient, who was known by the initials K.C., suffered extensive hippocampal damage as a consequence of a motorcycle accident. In addition to losing most of his episodic memories, he had a pronounced deficit in his ability to think about his own future. Below is an excerpt from a conversation between K.C. and Endel Tulving:

> ET: Let's try the question again about the future. What will
> you be doing tomorrow? [15 sec pause]
> KC: I don't know
> ET: Do you remember the question?
> KC: About what I'll be doing tomorrow?
> ET: Yes. How would you describe your state of mind when
> you try to think about it? [5 sec pause].
> KC: Blank, I guess.[6]

K.C. certainly understood the concepts of past, present, and future. He could order events in time, and he knew that his brother had passed away. K.C.'s deficit seems to be fairly restricted to what Suddendorf and Corballis would consider to be mental time travel. These and other observations are consistent with the notion that mentally traveling backward or forward in time relies in part on the same cognitive capacities we use to store and reconstruct autobiographical information about the past.

MENTAL TIME TRAVEL IN ANIMALS

Is the ability to mentally project oneself into the past or future unique to *Homo sapiens*? We have seen that all animals tell time and naturally anticipate external events: they learn to salivate in response to the bell before the food arrives, and can awake before the sun rises in order to set out in search of food. We also know that some animals seem to deliberately prepare for the future: birds build nests, beavers construct dams to protect their lodges, and squirrels store nuts. But do any of these behaviors imply that animals are in some sense thinking of the future, or that they grasp the concept of time?

Telling time certainly does not equate to thinking about the future; a clock tells time, it does not understand it. Furthermore, the acts of building nests or caching food do not imply that an animal understands the long-term consequences of its actions. No one would suggest that as a caterpillar searches for an ideal spot to anchor itself and become a pupa, it is thinking to itself "this is the perfect spot to transform myself into a beautiful butterfly." Most examples of apparent long-term planning in animals actually seem to be hardwired instincts. As the psychologist Daniel Gilbert has stated, "The squirrel that stashes a nut in my yard 'knows' about the future in approximately the same way that a falling rock 'knows' about the law of gravity"[7]—indeed, young squirrels that have never experienced a winter will stash nuts nevertheless. Animals perform all sorts of behaviors with no understanding as to why they are engaging in them, or of their long-term importance. Even humans have been known to engage in fairly complex behaviors with little thought as to what will happen nine months into the future.

But the fact that many future-oriented behaviors in animals are hardwired does not mean that animals *cannot* engage in mental time travel. Indeed, whether they do or do not is a hotly debated question in the fields of animal cognition and evolutionary psychology.

One of the leading candidates for animals capable of mental time travel is birds of the corvid family (jays, crows, and ravens). The British psychologist Nicola Clayton has led much of the research aimed at determining whether a species of jay, *scrub jays*, can engage in future-oriented mental time travel. Scrub jays cache small amounts of food in spatially distributed locations, and their excellent spatial memory allows them to recover the cached food at later times. As mentioned, in and of itself such caching does not imply mental time travel, but Clayton used a number of clever manipulations that together suggest the birds are doing much more than following their food-stashing instincts. Scrub jays will eat both worms (moth larvae) and nuts, but they prefer worms, at least when they are fresh. For instance, when given the choice between fresh worms and nuts, they'll take the worms, but when presented with nuts and five-day-old decaying worms, they demur at the worms and go for the nuts. So the question is, if scrub jays are allowed to cache both fresh worms and nuts, but only allowed to return to their caches either four hours *or* five days later, what will they choose to retrieve—the worms or the nuts? After a four-hour delay the birds were much more likely to search the locations they had placed the worms, but preferentially searched the nut locations when they were only allowed to return five days later (to ensure the birds were not basing their choices on odors emanating from the hiding places, the investigators always stole the cached food before the retrieval phase of these experiments). For example, in the four-hour group 83 percent of the first retrieval events were aimed at the worm locations, whereas this value was 0 percent in the five-day group. The birds seemed to realize the worms would have reached their expiration date after five days.

In another experiment, Clayton and her colleagues took advantage of the fact that scrub jays are known to engage in criminal activities. Jays that have seen another jay cache food may later steal it. As a countermeasure scrub jays are known to recache their food: if they know they have been observed during caching, they may later retrieve the food, not to eat it, but to rehide it. Clayton and colleagues showed that if a scrub jay knew it was being watched, it was more likely to come back and recache the treats compared to when it cached in private. This again suggests mental time travel, as one might argue that the jays are anticipating a future episode in which a crook steals their food. These and other related experiments have led Clayton to suggest that these birds are engaging in true forward-oriented mental time travel.[8]

Scrub jays are not the only serious candidates for mental time travel in animals. Other studies have asked if great apes (chimpanzees, bonobos, gorillas, and orangutans) exhibit what we would call foresight. One approach toward answering this question is essentially to determine whether great apes could get the hang of using money. One study was performed with apes that had previously learned to exchange tokens for food: they learned that trainers would accept certain valuable tokens, such as a piece of a colored PVC pipe, for food—other tokens, however, had no value. In the "foresight" experiments, apes also learned that after being given access to a bunch of tokens, thirty minutes later they would be given the chance to exchange the valuable tokens for food. So the experimenters asked how many of these tokens the apes grabbed before they were moved to a waiting room for thirty minutes. Six of the eight animals tested brought more valuable tokens to the waiting room compared to the number they brought during a control condition in which they could not exchange the tokens. The orangutans seemed to perform better than the bonobos, which in turn were better than the chimps. Overall it seems that at least some great apes have enough foresight to grab their billfold before going off to the market[9] (thus surpassing me on certain days).

Some scientists are not convinced that these results establish that apes are engaging in mental time travel. Perhaps the apes are mindlessly learning a sequence of actions, a much more complex version of rats learning to press a lever to obtain food. Furthermore, the effects were often weak; for example, it was generally the case that not every individual in the bird and ape studies "got it." Nevertheless, these studies provide compelling data that some animals can flexibly adjust their behavior to satisfy future needs. But the debate of whether animals are capable of true mental time travel will no doubt continue until there is a universally accepted definition or tests of mental time travel.

LIVING IN THE PRESENT

Regardless of whether or not our nearest living relatives are also capable of mental time travel, there is no doubt a gulf between humans and apes when it comes to thinking about and planning for the future. The primatologist Jane Goodall has stated "Chimps can learn sign language, but in the wild, so far as we know, they are unable to communicate about things that aren't present. They can't teach what happened 100 years ago, or ten years ago, except by showing fear in certain places. They certainly can't plan for five years ahead."[10]

Humans not only communicate about past events and make plans for the future, but hop back and forth on a mental time line to express complex temporal relationships. Consider the sentence: *Last month a preacher predicted the world will end in three months, so I will be spending all my savings next month.* Without the luxury of language and the ability to perform simple forms of arithmetic how could any animal make sense of such temporally complex ideas?

Some evidence for an interdependence between language, numbers, and mental time travel comes from studies of a remote hunter-gatherer tribe native to the Amazon: the Pirahã (pronounced *peed'a-han*). Their language has tenses for the simple past and future, but not the grammatical structure to express embedded temporal relationships such as *by next month I will have spent all my savings* (here the future perfect tense "will have spent" refers to the past from the perspective of some point in the future).

Numerically, the Pirahãs have a *one-two-many* system of counting: quantities above two are simply referred to as "many." They can discriminate between small and large numbers of items such as five and ten, but struggle to match the number of items between two different groups of objects. If shown four AA batteries and asked to place the same number of nuts as there are batteries on the table, they will do so fairly accurately, but they will generally fail at this task if there are ten items. Not surprisingly, given their one-two-many number system, they do not seem to have much of a notion of their age.[11]

The linguist and ex-missionary Daniel Everett believes the Pirahãs are grounded in the present: "Pirahãs don't store food, they don't plan more than one day at a time, they don't talk about the distant future or the distant past—they seem to focus primarily on the now."[12] Everett originally set out to learn their language, translate the Bible into Pirahã, and convert them to Christianity. He became fluent in their language, but failed epically in his missionary aspirations as the Pirahãs eventually led him to atheism. He thinks part of the failure was due to their lack of interest, and skepticism, toward events that they did not directly experience, or at least have secondhand knowledge of: they had little interest in stories of Jesus once they realized Everett had never actually met Jesus. Similarly, they seemed not to be preoccupied with the future, or what, if anything, happens after they die. Everett does not believe that the restricted temporal outlook of the Pirahãs reflects any sort of inherited neurological deficit, as they are intelligent and exquisitely skilled at surviving in the jungle: "They can walk into the jungle naked with no tools or weapons, and walk out three days later with baskets of fruit, nuts, and small game."[13] Rather, he thinks, the Pirahãs' present-based existence is a signature of their culture. Such moment-by-moment existence is certainly enabled by their environment and the more or less continuous availability of food. Their indifference towards the future would not be conducive to survival in indigenous Inuit cultures, where a significant amount of forethought and preparation go into surviving harsh winters.

SENDING MESSAGES INTO THE FUTURE

Different people and cultures vary dramatically in how much thought and effort they apply towards the future, and how far ahead they mentally travel into the future. We all know people who, a bit like the Pirahã, seem to live day by day—they are the ones who generally appear to be content, despite seeming to run into more than their share of financial and personal difficulties. At the other end of the spectrum are those whose every thought and action is aimed at achieving some goal in the distant future.[14]

And then there are the visionaries who dream decades and centuries into the future. This ability to time-travel beyond the lifespan of any individual is perhaps the cornerstone of human culture. Through folktales, cave paintings, stone and wood tablets, and eventually through writings on papyri and in books, *Homo sapiens* have engaged in a one-way conversation with future generations.

The 2004 Indian Ocean tsunami killed 230,000 people along coastal communities in fourteen different countries. One island community in Thailand, inhabited by the indigenous Moka people ("sea gypsies") was destroyed, but experienced few, if any, casualties. The elders knew of stories about the hungry spirits of the sea. They believed the receding of the sea (which precedes a tsunami) was a warning of the sea's hunger, this belief led the Moka to run to higher ground before the megawaves hit shore. To them the tsunami occurred because "the big wave had not eaten anyone for a long time, and it wanted to taste them again."[15] The stories told by survivors of past tsunamis must have been passed across centuries and stored not as boring facts in semantic memory ("when the sea recedes, run to higher ground"), but as visually rich and emotionally engaging stories about being eaten by the sea—and thus well suited to be vicariously stored in their episodic memory banks.

The semantic and episodic memories stored within our neural circuits are ultimately a recipe for survival. But the memories of any given human being are of limited capacity and accuracy, and are eventually erased altogether. Mental time travel has allowed us to see that future generations may benefit from these memories, and to create external storage devices, which can be used to pass down knowledge between individuals who will never stand face to face. Without such future-driven behaviors and cross-generational memories, modern culture, technology, and science would not exist.

TEMPORAL MYOPIA

Humans are the only creatures on the planet that can think about and plan for the distant future. We alone plant seeds that can take years to bear fruit or build structures to last across the centuries. And yet, many of the most serious problems facing modern man (and other species) are a consequence of human shortsightedness.

On the personal level a myriad of financial and health problems are related to our temporal myopia. Financial difficulties such as credit-card debt and retirement shortfalls, are often the consequence of shortsighted actions: either spending money we don't have or failing to save the money we know we will one day need.[16] Additionally, we often succumb to the short-term gratification of an unhealthy diet, or fail to exercise regularly at the expense of our long-term well-being. At the societal level, economic turmoil is often a consequence of the same flaws that result in personal financial troubles. Like individuals, governments often can't delay gratification or implement short-term sacrifices, sometimes choosing to enter further into debt rather than increase taxes or cut costs. Unsustainable debt, in turn, results in economic meltdowns that have profound long-term consequences, including unemployment and collapsing pension plans. Even in the absence of economic meltdowns pension funds are chronically underfunded; the reasons are multifaceted, but in the end they all underline Mark Twain's adage: "Never put off till tomorrow what you can do the day after tomorrow."[17]

A notable symptom of our temporal myopia at the societal level may be climate change. Even once we see the long-term consequences of our actions on the health of the planet, it is a challenge to take action. Despite our ability to foresee the future, we often struggle to care about time frames that extend beyond our own lifespan.

Like the gambler who perpetually believes his next bet will solve all his long-term troubles, our short-term thinking creates a vicious cycle of shortsighted actions that further compound our long-term problems. Perhaps one of the most severe consequences of our temporal myopia is that it hampers the effectiveness of the democratic process itself. Imagine a scenario where a hundred out of a hundred economists agree that the solution for long-term economic health is to immediately raise taxes. Come voting time, who is more likely to win an election: a politician running on the advice of the economists or one running on a platform to cut taxes?

The truth is that even though humans are far better at long-term planning than all other animals, we are not particularly good at it. This should not come as a surprise. The human brain is the product of an evolutionary process that unfolded over hundreds of millions of years. So most of our neural baggage comes from animals that lived, cognitively speaking, in the immediate present. Consequently, as a species, humans are still learning to perfect our newly acquired skills to better balance the allure of immediate gratification with the benefits of delayed gratification.

Which of the following two options would you choose: receiving $1,000 dollars right now or $2,000 dollars one year from now? There is no correct or incorrect choice here—although your average economist would be compelled to point out that a 100 percent yield in a year is hard to beat. This question captures the classic *intertemporal* trade-off: the immediate gratification of a reward available immediately versus the delayed gratification of a larger reward. Such intertemporal decisions permeate our lives. Should I buy a newfangled TV today and pay the credit-card interest rates, or save for few months until I have the cash? Should I play one more video game or get back to work? Should I spend more to buy an ecofriendly car in order to make a tiny contribution to the well-being of future generations?

Temporal discounting refers to the fact that the subjective value of something decreases with time. Receiving $1,000 today is in some very real sense more "valuable" than receiving the same amount a year from now. There is a chance that I will not be alive a year from now, so receiving $1,000 a year from now may be of zero value to me. To use a more naturalistic example, for one of our savannah-dwelling ancestors the promise of a small immediate meal far outweighs that of a larger meal a full moon from now if there is a chance he will die from starvation in the meantime. Throughout most of evolution our ancestors lived in a highly uncertain world—one in which starvation, predation, and disease were perpetual threats. Under such precarious circumstances short-term survival takes precedence over the relative luxury of worrying about the future. It is no wonder that humans come wired with a strong bias toward immediate gratification.

The balance between immediate and delayed gratification can be quantified by asking people, as in the example above, to choose between immediate small rewards and delayed larger rewards. By manipulating the size of the rewards and delays involved it is possible to calculate someone's *temporal discounting rate* in a specific context. Not surprisingly, there is a lot of individual variability. For example, in one study some people were very patient, willing to wait six months to receive $25 in lieu of receiving $20 immediately; others were much more impulsive, opting for an immediate payoff of $20 today over $68 in a month.[18] Numerous studies have shown that temporal discounting rates as measured in these intertemporal monetary choice tasks are inversely correlated with health, financial stability, and a propensity for substance abuse.[19] That is, people who are more likely to choose small immediate rewards in lieu of larger delayed rewards are a bit more likely to have health or financial problems.

When given a choice between $100 now and $120 a month from now, most people choose the immediate option. Knowing this, what do you think people prefer when both options are postponed by the same amount, that is, between $100 in a month or $120 in two months? Logically, if someone chose the immediate $100 over the $120 in a month, they should also choose the $100 in a month over the $120 in two months—in both cases they are waiting an extra thirty days for $20. This is not the case, however.[20] When both options are placed in the future, people become more patient. It is not worth waiting thirty days to get an extra $20 if we get $100 now, but the wait becomes worth it if both rewards are placed in the future. In other words, we favor the immediate rewards not because we are loath to wait thirty days for an extra $20 but, predictably, because we really like getting stuff right now!

Our bias toward immediate gratification is often exploited by financial institutions and marketers. The use of credit cards, for example, places a veil between the act of buying something and the fact that we are relinquishing our hard-earned cash. Studies show that people are prone to spend more when paying with a credit card versus cash. In one study students were willing to pay twice as much for sports tickets if they had to pay with a credit card compared to cash.[21] Furthermore, credit-card reward programs can further coax us into debt, by providing immediate "rewards" (airline miles, points, or cash back) every time we make a purchase—*spend more, get more*! (Consumers of course are ultimately paying for these "rewards.")[22]

Many of the astounding scientific, technological, and cultural accomplishments our species has achieved are a result of our ability to engage in mental time travel and execute long-term plans. But many of our personal and societal failures reflect the fact that many of our decisions are guided by immediate gratification.[23] Fortunately, how we balance the trade-offs between short- and long-term outcomes is not hardwired into our genes. Delaying gratification and making optimal intertemporal decisions is a process that benefits immensely from practice, education, deliberation, and simply stopping to think about the future. Studies show, for example, that temporal discounting rates can be extended—shifted from impulsive to more patient decisions—by having people engage in mental time travel as they make decisions. In one study subjects were given a series of intertemporal choices (for example, *$20 now or $60 in a month*). On some trials choices were accompanied by a phrase such as *vacation in Paris*, meant to trigger mental imagery of future events. Participants were less impulsive—that is, they chose the larger delayed rewards more often—in the mental time travel trials compared to the control trials.[24] So mental time travel itself may offer one means to debug our propensity for short-term gratification.

MENTAL TIME TRAVEL IN THE BRAIN

What makes human beings uniquely capable of mental time travel? Is there something different about the neurons of human beings? Is it the size of our brains? Or perhaps humans have brain areas that are not present in other animals?

Neuroscientists would be hard-pressed to tell a mouse neuron apart from a human neuron by measuring the electrical activity of these cells. Similarly, under the microscope the neurons of all mammals look very much alike. The human brain, of course, does stand out because of its size, but it is not the largest of the animal kingdom. Not surprisingly, bigger animals tend to have larger brains, so elephants and whales have much larger brains than ours. When body mass is taken into account, and the ratio of brain to body mass is examined, humans still stand out, but again we do not hold the record. Small animals tend to have bigger brains in relation to the size of their bodies, so even mice have slightly larger brain-to-body ratios than humans. The record holder is the tiny tree shrew: its brain accounts for around 10 percent the weight of its less-than-half-pound body, while the number is around 2 percent in humans. When the appropriate adjustments are made to account for the fact that the relationship between brain and body weight is not linear—the so-called *encephalization quotient*—then humans hold the record. Given the weight of human body, the human brain is more than 7 times larger than what would be expected based on the relationship between brain and body mass across all vertebrates. Dolphins have an encephalization quotient of a bit over 5. Chimpanzees lag far behind at around 2.5, and mice weigh in at a mere 0.5.[25]

There is little doubt that brain size, as quantified by the encephalization quotient, contributes to the unique cognitive abilities of humans. But the relative size of specific brain areas, or specializations within particular brain areas, also plays an important role. For example, relative to the whole brain, the auditory cortex is of similar size in primates and rodents; other areas, however, are proportionally larger in primates. One such area is the prefrontal cortex.

The prefrontal cortex, which is located right behind the forehead, is a very-well-connected brain area—that is, it is well situated to listen in on, and influence, what is happening in many other brain areas. While the prefrontal cortex underwent a preferential expansion in primates, the relative size of the prefrontal cortex is not proportionally larger in humans than in great apes.[26] There is some evidence, however, that the prefrontal cortex of humans is distinct in other ways— for example, neurons in human prefrontal cortex seem to receive more synapses.[27]

So what does the prefrontal cortex do? People who have suffered lesions to the prefrontal cortex can seem entirely normal at first encounter Their motor skills are largely intact; they can understand speech and talk normally; yet depending on the precise location and extent of the lesion, they have distinct deficits in higher-order cognitive functions. These include alterations in short-term memory, personality, attention, decision making, and inhibiting socially inappropriate behaviors. While people with prefrontal lesions can follow instructions and perform many tasks normally, they struggle to execute plans that require multiple steps and flexibly adapt as the circumstances change.[28]

The prefrontal cortex also contributes to our ability to make long-term plans, delay gratification, and engage in mental time travel—so people with prefrontal lesions are not the type to be saving much for retirement. One study used the temporal discounting task to examine how people with lesions to the prefrontal cortex balance immediate and long-term rewards. Compared to healthy controls and people who suffered lesions to other parts of the brain, prefrontal-cortex patients were significantly more likely to choose smaller short-term rewards in lieu of larger delayed rewards.[29] Similarly, a number of brain-imaging studies indicate that the degree of activity in parts of the prefrontal cortex is correlated with how long people are willing to delay gratification in temporal discounting tasks.[30]

Brain-imaging studies of healthy humans also suggest that the prefrontal cortex contributes to our ability to engage in mental time travel. For example, when people where asked to imagine a future scenario based on the name of a person and place that they knew, activity in the prefrontal cortex was higher than when they were asked to simply create sentences with those same words. Furthermore, studies also suggest the prefrontal cortex is more active when people are asked to imagine potential future events compared to when they are recalling past episodes of their lives.[31]

While the prefrontal cortex is important for mental time travel, it would be naive to say that is where mental time travel happens. Attributing any particular task to a specific brain area is a bit like watching a soccer game and asking whose job it is to score goals—players on the defense or offense certainly have different roles, but scoring is a team effort and in the end anybody can score. Future-oriented mental time travel is a complicated task that requires the orchestration of a number of different cognitive functions, including accessing past episodic and semantic memories, using these memories to conjure future scenarios, understanding the difference between the past and future, and the ability to judge whether the simulated outcome is desirable or not. Additionally, it is not sufficient to simply imagine future scenarios: we must remember what we imagined—in other words we must learn from our mental simulations. If you are planning a camping trip, you want to draw upon your memories of previous trips in order to decide what equipment you should bring. You also want to extrapolate from these memories to simulate novel worst-case scenarios: *what would happen if I were to sprain my ankle or be bitten by a snake?* Once you've simulated these scenarios—and assuming you are still going camping—it is important to learn from these simulations and make the appropriate preparations in case one of these scenarios were to transpire.

Given the cognitive complexity of mental time travel, it is to be expected that it relies on a collection of different brain areas working in concert. Indeed, lesion and imaging studies implicate a number of different areas in mental time travel. As mentioned above, the amnesic patient K.C. struggled not only to recall past episodes of his life, but to think about what he might do in the future. K.C.'s primary brain lesion was to the temporal lobes, the structure that contains the hippocampus (remember, the *temporal* in *temporal lobe* does not refer to time, but the temples or the temporal bone of the skull, which is beside the ears). One study asked people with medial temporal-lobe lesions to imagine, and then describe, different potential future scenarios, such as what it would be like to win the lottery. Compared to healthy controls, amnesics with temporal-lobe lesions provided impoverished descriptions and relatively few details about what the experience would be like.[32]

As is often the case, complex cognitive tasks such as mental time travel do not rely on any single brain area, but on a support network of many different areas, each contributing in its own way. In the case of mental time travel the medial temporal lobes may provide access to a foundation of past experiences, whereas the prefrontal cortex might flexibly manipulate these memories to dream up and evaluate novel scenarios. Interestingly, one thing mental time travel may not explicitly require is the ability to tell time. Much as a calendar represents time but does not actually tell time (it is not a clock), the neural circuits responsible for mental time must represent the past, present, and future but don't necessarily need to actively be able to measure the passage of time.

The ability of animals to predict nature's cycles and anticipate the behavior of predators, prey, and mates alike was a powerful evolutionary adaptation. Mental time travel was the next step: it made merely anticipating events in the external world an outmoded technology. Mental time travel allowed humans to go from passively predicting the future to actively creating it. Not enough food? Create a future in which there is an abundance of food through agriculture. Not enough water for agriculture? Create dams, channels, and irrigation systems.

How did our ancestors acquire the ability to mentally project themselves into the past and future? Do we understand the concept of time because we are capable of mental time travel, or do we engage in mental time travel because we grasp the concepts of past, present, and future? Answers to these questions will not be forthcoming from animal studies. Whether we call it mental time travel or not, scrub jays and great apes do have the capacity to guide their present actions towards desirable future outcomes but, compared to humans, there is clearly a vast chasm in their capacity for future-oriented thinking. If for no other reason, it is difficult to plan for the days, months, and years ahead without a semantic understanding of days, months, and years, or the ability to grasp the concept of time. Mental time travel is a multidimensional cognitive trait. It is likely a product of numerous converging evolutionary steps, including semantic and episodic memory, language, a number sense, and the spatialization of time into a mental timeline.

As mentioned previously, mental time travel is both a gift and a curse. Our trips to the future generally take us to places that we deem to be superior to our current circumstances, and often serve to outright escape the present. But as emphasized in Eastern philosophies, traveling to the past or future can preclude us from embracing the here and now as a primary source of happiness and joy.[33] Daniel Everett alludes to this: "The Pirahã simply make the immediate their focus of concentration, and thereby, at a single stroke, they eliminate huge sources of worry, fear, and despair that plague so many of us in Western societies." But if living in the present provides a more carefree life, it also provides much less of it (the Pirahã's average life span is around forty-five years, not taking into account infant mortality).[34] Ensuring a continuous supply of food, providing permanent shelter, engaging in scientific and artistic endeavors, and preventing and curing diseases all require vast amounts of foresight and planning. And therein lies the paradox of mental time travel: it seems to be both the solution and the cause of all our troubles.

12:00: CONSCIOUSNESS: BINDING THE PAST AND THE FUTURE

There is a theory which states that if ever anyone discovers
exactly what the Universe is for and why it is here, it will
instantly disappear and be replaced by something
even more bizarre and inexplicable.

There is another which states that this has already happened.

—DOUGLAS ADAMS

What does a baby perceive when she first opens her eyes? To the extent that she sees anything, it is surely a jumble of unfocused patterns, lines, and halos, devoid of meaning and impossible to interpret. In contrast, as you and I look out into the world we see a coherent and stunning reconstruction of the world around us: waves crashing along a sandy beach, kingfishers diving into the water, and even our own reflection on the water's surface. We generally, mistakenly, take this reconstruction to be real. But at best what we experience is correlated with the external physical world. The colors we see, for example, are simply an interpretation of the wavelength of electromagnetic radiation, as arbitrary as the link between the letters in the alphabet and the sounds we have assigned them. At worst, we see a fiction imposed upon the mind from within: from the visions of someone suffering from schizophrenia, to drug-induced hallucinations, to the dreams we experience every night. And there is so much we don't see: the bacteria living on our skin, the invisible galaxies in the sky, the muon particles created in the atmosphere, and the infrared heat signatures of those around us.

The feeling of the passage of time—our perception of change—is also a mental construct. To the neuroscientist this construct is correlated with reality: we perceive waves crashing and birds diving into the water because *time is actually flowing*—these events are unfolding in a universe in which only the present is real. To many physicists and philosophers the flow of time is also a mental construct, but of something that holds no equivalent in the physical world. Within the block universe of eternalism our feeling of the passage of time is more akin to the visions of a schizophrenic, something that only exists within.

These two views offer incompatible notions of the nature of time, but they both consider our feeling of the passage of time to represent a fundamental problem. Resolving this problem, however, will prove to be a formidable task, as our subjective sense of time sits at the center of a perfect storm of unsolved scientific mysteries: consciousness, free will, relativity, quantum mechanics, and the nature of time.

SHARDS OF TIME

Let's assume for a moment that we live in a presentist universe in which only the now is real. Our conscious perception of time resonates with presentism, as consciousness seems to provide a continuous play-by-play report of the events unfolding around us. But this too is an illusion in the sense that while the unconscious brain continuously samples and processes information about events unfolding in time, consciousness itself is generated in a highly discontinuous manner. The unconscious brain delivers stories to the conscious mind in fits and starts.

As you listen to an actress in a play deliver her soliloquy, you do not consciously perceive the continuous flow of each syllable of her speech. Rather, the meaning of words and phrases materialize fully formed into your mind.

You can easily identify the syllable *po* when heard in isolation, yet you are not consciously aware of that syllable when you hear the word *hippocampus*, nor do you consciously become aware that the word *camp* is embedded within *hippocampus*. Clearly our brains are not providing a linear play-by-play account of the raw sensory events unfolding in the external world.

The temporal structure of consciousness is actually a highly edited version of reality. If you gaze into the eyes of a friend while asking her to move them back and forth, you can easily see these movements taking place over time. Now if you were to gaze deeply into your own eyes while looking in the mirror, and then proceed to move them back and forth you will find yourself unable to see your own eyes move. These rapid voluntary movements of the eyes are called saccades, and more rigorous experiments than trying to look at your own eyes in the mirror demonstrate that vision is partially suppressed during saccades.[1] For example, if an image is flashed onto a screen as you are moving your eyes to a new position, and removed before the eye movement has ended, it is likely that you will not consciously perceive the flashed image. As far as consciousness is concerned, the brain often deletes the frames that occur during eye movements. Similarly, every blink we take blanks out the raw input to the visual system; these blanks are not consciously perceived because the brain fills in the gaps by splicing the frames before and after the blink. According to one estimate, between saccades and blinks, a full hour of visual information is lost throughout the course of a day, without any perceived blanks in our visual stream of consciousness.[2]

RECALIBRATING TIME

Thunder and lightening are caused by the same event, but, with any luck, we perceive them separately because the speed of light is close to a million times faster than the speed of sound. The sound of thunder reaches our ears significantly after the photons generated by the lightning reach our eyes. In other instances, however, the brain must not only process the input streams from the eyes and ears in parallel, but also attempt to align and synch the inputs from both of these sensory modalities. As mentioned earlier, when orchestral cymbals clash, the light waves and sound waves arrive at the eyes and ears at different times, but they are not perceived as distinct events. Rather the sight and sound of the cymbals clashing are integrated into a unified, multimedia experience before being "delivered" to consciousness. The same is true of speech. When someone says the word *baby*, we see her lips come together and then open to release the syllables *ba* and then *by*. Again, the sight of the lips opening will arrive at the eyes a bit before the sound arrives at the ears. The delay can be significant: in a large classroom it may take fifty milliseconds for the professor's voice to reach the back of the room. Similarly, the sound of a bat hitting a baseball will be delayed just over 100 milliseconds for the shortstop. Yet we generally perceive a speaker's lip movements and speech (or the sight and sound of a bat hitting the ball) as unified events.

One might suspect that we generally don't register the delay between visual and auditory signals because the brain does not have the temporal resolution to detect differences of 50 or 100 milliseconds. Not so. With practice, people can detect delays of around 20 ms between the onset of two tones of different frequencies.[3] The reason we do not consciously register the delay between visual and auditory signals is that the unconscious brain does its best to deliver an integrated interpretation of events. The time span during which the brain integrates visual and auditory information into a single unified percept is called, appropriately enough, the *temporal window of integration*. Within this window, subjectively speaking, the brain considers the auditory and visual events to be simultaneous. The windows can be over 100 milliseconds for speech—for example, if there is a mismatch of less than 100 milliseconds between the audio and visual tracks of a movie, it rarely comes to our attention. But the window is asymmetric—that is, if the auditory signal *precedes* the visual signal by 50 milliseconds, subjects may notice something is awry, but not if the auditory *follows* the visual signal by 50 milliseconds.[4] Another indication that the brain is actively attempting to align signals from the auditory and visual modalities is that the temporal window of integration is not set in stone—it is not a consequence of a fixed delay in the visual system (although visual information does take more time to arrive in the cortex because the eye is slower than the ear[5]); rather, it is adaptive. Studies show that after seeing a few hundred light flashes, each followed 200 milliseconds later by an auditory tone, people may judge a subsequent flash and tone, separated by 20 milliseconds, to have occurred simultaneously. However, they might judge this same flash-tone pair to have not occurred simultaneously if they had just listened to hundreds of flashes *preceded* by tones. In other words, by consistently exposing people to artificially long visual-auditory delays, it is possible to shift or expand people's temporal window of integration—thus, subjective simultaneity is relative. Based on past experience, the brain accommodatingly creates a narrative in which those visual and auditory signals are now integrated into a single event.[6]

Perhaps the most compelling example of how consciousness reflects a temporally edited account of reality is that later sensory events can actually alter our conscious perception of earlier events. Speech again provides a good example. Consider listening to the sentences *The mouse broke* versus *The mouse died*. The meaning of the word *mouse* can only be established at the end of each sentence. Yet it is not generally the case that you become consciously aware of a computer mouse or a rodent, and then wait for the last word to interpret the sentence.

The standard example of this *backward editing in time* comes from the so-called *cutaneous rabbit illusion*. Imagine someone taps your forearm twice near your wrist, and then in rapid succession, taps two times again near your elbow. What people often report to have happened is not what actually happened. The perception is not of two taps near the wrist and then two more near the elbow, but rather of four taps hopping along your arm: starting from the wrist to the elbow with two points in between.[7] If someone taps you twice on the wrist, and leaves it at that, you'd correctly report feeling two taps on the wrist. But in the rabbit illusion the third and fourth taps alters the perceived localization of the second tap. The take-home message is that the location of later stimuli alters your conscious perception of the location of the earlier ones, thus consciousness cannot be a uniform and continuous account of the flow of time. Rather, it seems that the unconscious brain is continuously processing the input stream, but waits for critical junctures before sending a polished narrative into consciousness.[8]

CORRELATES OF CONSCIOUSNESS

We do not know how the brain performs these temporal feats, much less how the brain goes about generating consciousness. But progress has been made in attempting to identify some of the *neural correlates of consciousness*: the patterns of neural activity that may underlie conscious perception.[9] A typical experiment uses EEG recordings in which electrodes on the scalp pick up small electrical signals from the cortex. One strategy used to hunt for the neural correlates of consciousness is to compare the electrical activity generated by a stimulus that is consciously perceived with the signals generated when the same stimulus is only registered subliminally—that is, when the brain detects the stimulus, but it does not bubble into consciousness.

In the laboratory researchers can straddle the threshold of conscious perception by flashing a stimulus, such as a tilted line, for less than 100 ms in one quadrant of a computer screen. In one study, subjects were asked to quickly indicate in which quadrant they thought the stimulus was presented (guessing if necessary), and whether they actually saw the line—that is, did they consciously perceive it or were they guessing. In the trials the subjects reported that they were guessing, they should be correct approximately 25 percent of the time. Interestingly, however, participants were correct much more often, indicating that subjects were subliminally detecting the stimulus— in other words, the unconscious brain knew where the stimulus was, but did not bother to convey this information to the conscious mind. Now the question is, what is the difference between what is happening in the brain between the correct trials in which subjects reported seeing the stimulus (correct/aware) and correct trials in which the subjects were "guessing" (correct/unaware)? The electrical activity in the correct/aware and correct/unaware trials was essentially identical for the first 250 ms after the stimulus—so whether or not subjects were aware of the stimulus there was no detectable difference in brain activity. At around 300 ms, however, a clear increase in cortical activity was observed throughout the brain in the aware trials.[10] These and numerous other studies suggest that the neural mechanisms underlying the conscious perception of a stimulus only emerge significantly after the stimulus has been detected by the brain. As explained by the French neuroscientist Stanislas Dehaene: "Not only do we consciously perceive only a very small proportion of the sensory signals that bombard us, but when we do, it is with a time lag of at least one-third of a second. . . . the information that we attribute to the conscious 'present' is outdated by at least one-third of a second. The duration of this blind period may even exceed half a second when the input is so faint that it calls for a slow accumulation of evidence before crossing the threshold for conscious perception."[11]

So the lesson is not simply that consciousness provides a delayed narrative of the events taking place in the extracranial world, but that the delay is variable. If someone yells the word *fire*, it probably takes relatively little time to reach consciousness because the unconscious brain can quickly find a relevant narrative to send into consciousness. But when hearing *the top stopped spinning* or *the top of the mountain*, the unconscious brain presumably waits for an unambiguous interpretation of the word *top* before crafting a conscious percept.

The brain cuts, pauses, and pastes the reel of reality before feeding the mind a convenient narrative of the events unfolding in the world around us. Yet unless we stop to think about it, we are left with the impression that our conscious experiences reflect an instantaneous play-by-play account of reality.

TIME AND FREE WILL

Much like the word *time*, *free will* is one of those concepts that, to quote Saint Augustine again: "I know what it is. If I wish to explain it to him who asks, I do not know."

If a tree falls in a forest and no one is around to hear it, does it make a sound? The riddle here lies in the ambiguity of the word *sound*: if one correctly defines *sound* as the vibration of air molecules, then the fallen tree makes a sound, but if one were to define sound as the conscious perception of those vibrating molecules by a human, then the answer is *no*.

The riddle of whether free will exists also revolves around the ambiguity in defining *free will*.[12] One of the definitions of *free will* in the Oxford English Dictionary is "The power of an individual to make free choices, not determined by divine predestination, the laws of physical causality, fate, etc."[13] This is perhaps what most people mean by free will, but scientifically speaking it is an unfortunate definition, because if by "physical causality" one means the laws of physics, then one is only left with the possibility that free will is the product of some ineffable substance or entity akin to the soul. Indeed, as the neuroscientist Read Montague has put it, "Free will is the close cousin to the idea of the soul—the concept that 'you,' your thoughts and feelings, derive from an entity that is separate and distinct from the physical mechanisms that make up your body."[14] On the upside, adherence to this definition would allow us to actually answer the riddle: no, free will does not exist, as the concept of a soul is a creation of the human mind, not the source of the human mind.

Less restrictive definitions of free will run along the lines of: *the ability to choose between different possible courses of action*, or *the ability to act without suggestion or coercion*. But these are hopelessly vague definitions—they do not constrain whether choices must be conscious or not, and would seem to allow for the possibility that a computer chess program is exerting its free will every time it checkmates me. More helpful are definitions that equate free will with unpredictability. For example, Stephen Hawking states, "The reason we say that humans have free will is because we can't predict what they will do."[15] Similarly, for Roger Penrose, "The issue of free will is discussed in relation to determinism."[16] In other words, if the laws of physics establish that it is possible to predict the state of any system at time *t*, including the human brain, from previous moments in time, then free will does not exist. As explained by the philosopher Michael Lockwood: "Universal determinism is so widely held to be incompatible with the existence of free will. For universal determinism is the thesis that the universe is subject to a set of rigid laws that, in conjunction with the state of the universe at any given time, prescribe precisely what state the universe will be in at any subsequent time. If the universe really is deterministic in this sense, it follows, . . . that all future outcomes—including, therefore, all our own future choices and actions—are already fixed."[17]

In this context, quantum mechanics is a thorny theory because, unlike the rest of physics that deals in certainties, quantum mechanics deals in probabilities. We know that at some level quantum events must affect the state of the brain—after all, every photon detected (or not) by your retina is playing by the probabilistic rules of quantum mechanics. So even in theory it is probably impossible to predict human behavior with 100 percent accuracy. Nevertheless, quantum mechanics provides a form of probabilistic determinism: it establishes a domain of options and their respective probabilities, leaving in the eyes of many philosophers little room for free will.

But for those who believe we live in a 4D block universe, the issue of whether the laws of physics are deterministic or not is rather secondary, as the block universe itself leaves no room for free will. If the past, present, and future all coexist within the block universe, all choices to be made have "already" been made.

All of the above definitions leave unaddressed one fundamental aspect of free will: the irrepressible feeling that we are in control of our own choices. We can debate whether or not we are actually "free to choose," but we can agree that it certainly *feels* like we are.[18] So perhaps we should define free will as exactly that: a feeling. As the psychologist Daniel Wegner defined it in the early aughts, free will "is merely a feeling that occurs to a person. It is to action as the experience of pain is to the bodily changes that result from painful stimulation."[19] Defining free will as the flavor of consciousness associated with the neural processes responsible for making decisions is not a new idea. Almost three hundred years ago the philosopher David Hume stated that "by the will, I mean nothing but the internal impression we feel and are conscious of, when we knowingly give rise to any new motion of our body, or new perception of our mind."[20] Thomas Huxley also argued that "the feeling we call volition is not the cause of a voluntary act, but the symbol of that state of the brain which is the immediate cause of that act. We are conscious automata, endowed with free will in the only intelligible sense of that much-abused term."[21]

ARE HUMANS PREDICTABLE?

If we choose to define free will as the feeling that occurs after the brain makes a decision—that is, after the unconscious neural processes within the brain make a decision—it should be possible to detect neural signatures of those decisions before people are conscious of them. There are numerous studies that suggest this is indeed possible. These types of experiments are only feasible as a result of one of the standard treatments of severe epilepsy. Patients who suffer from intractable epilepsy sometimes undergo a surgical procedure aimed at removing the part of the brain responsible for triggering the seizures. To precisely locate the epileptic focus, neurosurgeons implant electrodes in the brain and wait until the patient has a seizure. By recording the neural activity during a seizure, they are able to accurately target the pathological area for surgical removal. In the name of science, patients often agree to participate in experiments while the electrodes are implanted and doctors are awaiting seizure activity. In a study led by the UCLA neurosurgeon Itzhak Fried, electrodes were implanted in an area of the frontal lobe called the *supplementary motor area*. The patients were asked to perform a very simple task: to exercise their "free will" by pressing a computer key whenever they wanted too. Many neurons changed their level of activity well before the button was pressed. Indeed, based on the activity of a population of neurons it was possible to predict that a patient was about to move a finger with an accuracy of over 80 percent, a full 900 ms before the key was pressed (and 700 ms before the patients reported being aware of having decided to move).[22] Note that 900 ms is more than enough time for the brain to execute a finger movement; for example, if you were asked to press a computer key as soon as you saw a flash of light, the delay between the flash and the key press would be around 300 ms. Amazingly, these studies suggest that it was possible for the experimenters to know the person was about to press the key before the volunteer was aware he was going to press the key.

A series of related studies has confirmed that, at least under some circumstances, it is possible to determine what actions humans and animals will take hundreds of milliseconds, or even full seconds, before the action is taken.[23] But these results do not necessarily mean that it is possible to accurately predict human behavior from patterns of neural activity, or that consciousness does not contribute to our decisions. The decisions being made in these studies are very simplistic. Deciding when to move your finger does not compare to deciding whether or not to accept a job offer. So while it may be the case that choosing when to flex your finger is determined by unconscious processes that trigger the conscious feeling of free will, the participant's decision to participate in the experiment in the first place likely depends on a mixture of unconscious and conscious neural processes.

How does the computer key come to be pressed? Voluntarily pressing a key requires contraction of the muscles of the finger, which requires a barrage of action potentials traveling down the median nerve, which requires activation of motor neurons at the cervical level of the spinal cord, which is triggered by activity in the area of the motor cortex that maps on to the hand. But what causes those motor cortex neurons to fire? Well, now things start to get very messy, but the bottom line is that triggering activity in any neuron usually requires a surge of activity in the neurons synapsing onto that neuron (its presynaptic partners). Loosely speaking, this surge can take the form of many presynaptic neurons firing within a short time window of a few milliseconds (this is called *spatial summation*, which is a bit like quickly filling a tub by simultaneously dumping many buckets of water into it), or by a few presynaptic neurons firing continuously over a time window of tens of milliseconds or longer (this is called *temporal summation*, like using a single faucet to fill the tub). Either way we can think of this as a *progressive accumulation of drive or evidence* toward a given decision—you may decide to see a movie because on one occasion many of your friends told you to go see it (spatial summation), or because one very insistent friend told to you see it on many different occasions (temporal summation). Another factoid relevant to this discussion is that neurons are "noisy": their activity spontaneously fluctuates, going up and down without any apparent reason (there are, of course, causes for these fluctuations, but we will simply attribute them to random background noise). We can envision a simple decision to press a button or not as a "race" between two groups of neurons—say those cortical motor neurons responsible for lowering the finger onto the computer key versus those responsible for raising it. One group of neurons may get a head start because of random fluctuations, and on any given trial, the "willed" decision to press the key or not could be triggered by unconscious and random fluctuations within specific circuits in the brain—once a group of neurons wins the race, a motor movement and the conscious feeling of free will is generated. One explanation as to why neuroscientists can predict the movement of a patient's finger hundreds of milliseconds in advance is because they are picking up which group of neurons gets off to a head start.

It's too early to draw definitive conclusions from neurophysiological studies of free will, but in a field famously devoid of experimental data, these experiments serve as an anchor for all discussions of free will. And as summarized by the neuroscientist Patrick Haggard, it is becoming increasingly accepted that "although we may experience that our conscious decisions and thoughts cause our actions, these experiences are in fact based on readouts of brain activity in a network of brain areas that control voluntary action."[24] What we consciously perceive as free will is presumably preceded by unconscious neural computations that are responsible for making decisions. Indeed, it is hard to see how it could be otherwise. Everything we know about the brain is consistent with the fact that all mental states are produced by a pattern of neural activity within the brain, and any given neural pattern is produced by the interaction of the previous neural state (both the active and hidden states discussed in chapter 6), the current external input, and stochastic fluctuations occurring at the thermodynamic and quantum levels.

The notion that free will is merely a feeling that occurs after consciously inaccessible circuits make a decision can be unsettling. Indeed, it has been argued that if this were the case, then consciousness would be useless, "the proverbial backseat driver, a useless observer of actions that forever lie beyond its control."[25] But even if consciousness, along with the feeling of free will, is an after-the-fact mental creation, it doesn't follow that consciousness doesn't play a role in decision making! If you go on a blind date and during dinner your companion suddenly picks up a fork and sticks it into your hand, the rapid withdrawal of your hand is probably too quick to be attributed to consciousness—thus your action was likely independent of your conscious perception of pain. But your conscious perception of the pain likely influences subsequent decisions, such as whether or not to go on that second date. Evolutionarily speaking, subjective experiences and free will may be primarily future-oriented phenomena. For example, perhaps it is the feeling of free will that provides the conviction that we are in control of our destiny, and thus the impetus to take charge and make the long-term, future-oriented, actions necessary for survival.

CRIME AND PUNISHMENT

The debate over whether or not free will exists is central to questions pertaining to moral responsibility and the justice system.[26] Some argue that if our decisions stem from deterministic and unconscious processes within our neural circuits, we would not be responsible for our own actions—in other words, that determinism is incompatible with moral responsibility. For example, the physicist George Ellis, a proponent of the evolving block universe (in which the past is a frozen 4D spacetime block, but the future does not yet exist), believes that eternalism prevents the expression of moral responsibility: "If we are just machines living out a future that has already been set, then Adolph Hitler had no choice to do other than what he did. . . . To me, that's an untenable view that will lead to great evil because people will just stand by as evil takes place."[27] In agreement with Ellis's concerns, surveys suggest that when people are told that all our decisions are a consequence of deterministic and unconscious events, they are less likely to hold others accountable for their actions.[28]

Let's consider three cases in which a pedestrian is injured by a motorist: (1) the driver of the car had carefully devised a plan to hit the pedestrian; (2) the driver lost control of the car while checking his text messages; and (3) the driver lost control of his car while in the midst of suffering his first epileptic seizure.[29] Because all three scenarios can ultimately be traced to the murky inner workings of the motorist's brain, some might be concerned that in the absence of what most people think of as free will, in all three scenarios the motorist "had no choice to do other than what he did." This concern, however, is in part an atavism of the belief in a soul, a form of *cryptodualism*—in which we implicitly assume that the mind is independent of the brain. If I make the decision to check my text messages while driving, does it matter if the key neural events that triggered that decision were unconscious or conscious, predictable or unpredictable, or predetermined or not? The decision was made by my brain, that is, by me—there is no distinction between me and my brain! This is not to say that the three scenarios above are in any way equivalent, or that the mental state of the driver is not relevant when meting out punishments. But we should not confuse the question of punishment with that of responsibility. In all the scenarios above, the driver is responsible—regardless of whether decisions are made consciously or unconsciously. Indeed, this is currently reflected in the justice system: the motorist will be held responsible in all three cases (for example, he will be liable for medical bills of the pedestrian). However, the punishments in each of the above cases will rightly differ, as punishments should take into account a complex set of factors, including whether a crime was premeditated (which establishes an intent to do harm—whether or not intent emerges from unconscious or conscious process), and likelihood of future offenses and rehabilitation.

Neuroscientists, physicists, philosophers, and legal experts will continue to debate questions pertaining to moral responsibility, determinism, and the role of conscious and unconscious processes in decision making. But perhaps it is time to use our "free will" to embrace the notion that free will is the conscious feeling associated with the neural processes underlying our decisions, decisions that we are fully responsible for because each and everyone of us is the sum of our unconscious and conscious selves.

———

There are few experiences more persuasive than the feeling of each fleeting *now* vanishing into the past while opening the doors to an endless array of potential futures. It is because this feeling is so compelling that the concept of eternalism comes as an assault on our grasp of reality. Yet, counterintuitive as it may be, the notion that the past and future are as equally real as the present is the favored theory about the nature of time. But this block-universe view is not without failings. Indeed, there is no unified consensus about the nature of time, as it is widely recognized that time plays distinct roles within the laws of physics. Ongoing efforts in physics, for example, are aimed at solving what is known as the *problem of time*: the conflict between the role of time in general relativity and quantum mechanics. In general relativity, time (as a constituent of spacetime) can be thought of as part of the fabric of the universe, whereas in quantum mechanics time is a parameter that governs the evolution of a quantum system. Puzzlingly, however, some mathematical attempts to merge general relativity and quantum mechanics end up without any role for time whatsoever. The time parameter simply disappears from the equations,[30] leaving one with the impression that we actually live in a block universe composed only of three spatial dimensions.

The challenges to eternalism and the block universe do not only arise from within physics, but from neuroscience as well. Most notably, the block universe fails to explain the fact that we perceive time to be flowing, leading some to suggest that the subjective experience of the passage of time is a mental artifact. Could one of our most salient and universal experiences be an illusion in the deepest sense of the word *illusion*? The geneticist Theodosius Dobzhansky famously noted that "nothing in biology makes sense except in the light of evolution." If we are to take this statement seriously, it stands to reason that what we are, and are not, conscious of is a product of evolutionary pressures.[31] So presumably some functions the brain performs generate subjective experiences because consciousness affords a selective advantage to these processes. We are conscious of painful stimuli because the subjective experience of pain must offer a selective advantage not afforded by zombie-like reactions to injury (perhaps the advantage lies in protection from future injuries). By this reasoning, the feeling of the passage of time should also provide a selective advantage. But what could this advantage be if we live in the frozen block universe of eternalism?

A further conflict between physics and neuroscience is that if the flow of time is an illusion created by the mind, then instantaneous slices of the block universe must be able to sustain the phenomenon of consciousness. Yet we are not conscious of instantaneous moments, but rather of chunks of time that capture meaningful and interpretable events—the "specious present". Even more vexing is the question of whether the phenomenon of consciousness itself requires some temporal thickness. Perhaps consciousness is more akin to evolution, an inherently temporal process that cannot really be said to exist within a static frame.

As physicists and philosophers continue to grapple with the problems of time within physics, the neuroscience of our perception of time's flow should be part of the debate. We must determine whether our subjective sense of the passage of time reflects a physical phenomenon that needs to be explained by physics, or rather a rare subjective experience that does not correlate in any way with reality. Neuroscientists and psychologists in turn must acknowledge the fact that the brain is at its core a temporal organ. If one were to unwisely attempt to summarize the function of the brain in three words, those words might be *anticipate the future.* The brain tells time, generates temporal patterns, remembers the past, and endows us with the ability to mentally project ourselves forwards in time—all in order to predict and prepare for the future.

Because time is so critical to brain function, telling time is built into the neural operating system at the deepest levels: synapses, neurons, and circuits. It makes no sense to ask which part of the brain tells time because most of the brain's circuits tell time in one form or another. The multiple clock principle tells us that, unlike a wristwatch that can track milliseconds to years, the brain has an array of distinct mechanisms to tell time across different scales—and even within a given temporal range, different circuits are responsible for timing, depending on the task at hand. Why are the clocks within the human brain so radically different from those devised by the human brain? The answer lies in part in the building blocks of man-made and neural clocks. Man-made clocks rely on the counting of each consecutive tick of an oscillator—the faster the period of the oscillator, the more ticks to count. The building blocks of the brain lack the precision and numerical range of the digital components of modern clocks—neurons cannot count to 32,768, much less 9,192,631,770.

Telling time is a skill we share with all animals, but what makes *Homo sapiens* unique is the ability to transcend nature's capricious ways by peering into the future and shaping it to meet our needs. But mental time travel is a gift and a curse. In peering into the future our ancestors must have foreseen more than they were prepared to cope with: their own inevitable death. This disturbing vision perhaps led them even farther into the future, and to the invention of extreme mental time travel: they envisioned an afterlife.

Mental time travel requires a delicate balance of science and art: rigorous extrapolation of the remembered past and dreaming up the inconceivable. This balance can go awry. We sometimes spend too much time dreaming, thus failing to address scenarios that we are perfectly capable of foreseeing and preventing. Today we struggle to recognize and address the long-term economic, health, and environmental problems that face our species. This temporal myopia is understandable: evolutionarily speaking, mental time travel is a newly acquired skill. Fortunately, like other cognitive skills, mental time travel benefits immensely from practice and education.

Perhaps we live in a universe in which only the *now* is real, or maybe the *now* is as arbitrary as the *here*. Or maybe the nature of time is even more bizarre and inexplicable than anything we have yet conceived. But regardless of the true nature of time, there is no excuse for not continuing to hone our mental time-travel skills. To learn to better discriminate the improbable from the impossible, to embrace long-term rewards over short-term gratification, and to act in the *now* to create a future that we'll want to inhabit when it eventually "becomes" the present.

ACKNOWLEDGMENTS

Clock time is the ultimate equalizer. Each of us is given 86,400 seconds every day. Innumerable books have been written to help us best use this daily allotment of seconds. The psychologist Phil Zimbardo suggests that we view our parcels of time, particularly so-called free time, as packages to be wrapped and gifted to the people and activities we value the most. I am grateful for the many friends and colleagues who have given me the gift of their time, for without them this book would not have been possible.

This book is the product of a long-standing interest in how the brain tells time—a problem I have dedicated much of my scientific career to. Throughout my voyage I have benefited immensely from the work and encouragement of a number of leaders in the timing field, and I'm particularly grateful for the support of Richard Ivry, Michael Mauk, and Warren Meck. I am also thankful for the teaching and wisdom of my other friends in the field: Domenica Bueti, Catalin Buhusi, Jenny Coull, Mehrdad Jazayeri, Hugo Merchant, Matt Matell, Kia Nobre, Virginie van Wassenhove, and Beverly Wright.

Part of the reward of writing a book is the joy of exploring fields outside the comfort of one's narrow scientific niche. In writing this book I made the decision to dip a toe in the physics of time, and I am indebted to a number of physicists and philosophers who patiently answered my naive questions about time, relativity, and quantum mechanics, including Richard Arthur, Vincent Buonomano,

Sean Carroll, Craig Callender, Per Kraus, Dennis Lehmkuhl, Terry Sejnowski, Lee Smolin, and especially Harvey Brown.

I am no doubt guilty of omitting and oversimplifying the contributions of many colleagues and scientists in the name of brevity, while attempting to write an accessible account of a vast scientific field. But in the end these failings are likely more a symptom of my own shortcomings as a scientist and writer. I thus apologize in advance to those scientists whose work I have omitted or not properly credited.

Many friends, colleagues, and collaborators either read individual chapters or educated me on some of the material covered in this book. In particular I'd like to thank Judy Buonomano, David Burr, Chris Colwell, Jack Feldman, Paul Frankland, Dan Goldreich, Jason Goldsmith, Vishwa Goudar, Sam Harris, Nicholas Hardy, Sheena Josselyn, Rodrigo Laje, Michael Long, Hakwan Lau, Helen Motanis, Joe Pieroni, Carlos Portera-Cailliau, Rafael Núñez, and Alcino Silva.

My own research on how the brain tells time would not have been possible without the support of the National Institute of Mental Health and the National Science Foundation, as well as the Departments of Neurobiology and Psychology at UCLA. I am grateful to Annaka Harris and my editor at Norton, Tom Mayer, for their guidance and editorial expertise. Finally, I'd like to thank my wife, Ana, for giving me the gift of so much of her time.

NOTES

1:00: FLAVORS OF TIME

1   http://oxforddictionaries.com/words/the-oec-facts-about-the-language.

2   More precisely, we see objects distributed in space.

3   The retina of some species does have cells that detect motion—that is, whether an object is "moving" in time (and space). Additionally, it should be pointed out, the cochlea does in a sense tell time, because the sensory cells of the cochlea (the inner hair cells) are tuned to the frequency of the vibration of air molecules—and frequency is a measure of the time it takes an oscillation to complete a full cycle. These frequencies are much too fast for most neurons to be responsive to, and much too fast to allow for conscious perception of these intervals.

4   From the Oxford English Dictionary.

5   For a detailed discussion of the history of the pendulum see Matthews, 2000.

6   A number of popular science books provide an excellent overview of the progressive steps in the history of mathematics and physics, such as Penrose, 1989.

7   Barbour, 1999.

8   Wells, 1860.

9   I am oversimplifying here, there have been a few highly influential books and articles published in the mid-twentieth century that to a certain extent withstood the test of time—e.g., works by Lashley (1951) and Fraisse (1963).

10 Kandel et al., 2013. You will also not find the word *timing* but will find the word *temporal*, but most of these entries pertain to the *temporal lobe* (*temporal* can refer either to the temple or to that which pertains to time—which has led more than a few people to assume that our ability to tell time is located in the temporal lobe). Note that I use this example not to suggest there is an omission within this textbook, but as a representative example of the relative neglect of the problems of time in neuroscience in general.

11 Ivry and Schlerf, 2008.

12 Dudai and Carruthers, 2005; Tulving, 2005; Schacter and Addis, 2007; Schacter et al., 2007.

13 Most philosophers and physicists would agree that building a time machine would provide a strong argument against presentism. In the current context I am essentially stipulating that time travel (more specifically, closed time-like curves in spacetime) are incompatible with the notion of presentism as I am using the term. But there are ambiguous cases. For example, one might argue that circular time, in which the present loops back on itself, is compatible with both presentism and a form of time travel to the past (generally speaking this is not what we mean by time travel). But in general, as stated by the philosopher Michael Lockwood, "time travel and the tensed, common-sense view of time . . . simply do not mix. The very concept of time travel makes sense only in the context of a tenseless view of time" (Lockwood, 2005). Note that Lockwood is using the terms *tensed* and *tenseless* (*untensed*) time as I use the terms *presentism* and *eternalism* respectively.

14 Quoted from Davies, 1995, 253. See also Smart, 1964.

15 Weyl, 1949/2009.

16 Einstein, 1905.

2:00: THE BEST TIME MACHINE YOU'LL EVER OWN

1 Apparently there are a few true time-travel antecedents to *The Time Machine*, including the Spanish author Enrique Gaspard's book *El Anacronópete*. I should stress that my knowledge of the literature is highly limited, and I certainly did not conduct an exhaustive search of the history of time travel in fiction. So there are likely some exceptions to the statement that true time travel only emerged in the late nineteenth century.

2 Even if the past and future are as real as the present, and the laws of physics do not explicitly prohibit time travel, they may very well conspire to ensure that it is impossible in practice—a notion that Stephen Hawking has called the "Chronology Protection Conjecture." Many excellent popular-science books and articles have been written about the possibility and physics of time travel, including: Davies, 1995; Thorne, 1995; Carroll, 2010; Davies, 2012.

3 Dennett, 1991, 177; Clark, 2013.

4 Henderson et al., 2006.

5 Tulving, 2005.

6 Hume, 1739/2000), 116.

7 Földiák, 1991; Wiskott and Sejnowski, 2002; DiCarlo and Cox, 2007.

8 In the context of classical conditioning there is an exception. Humans and other animals will develop conditioned taste aversion between eating a food

and becoming sick, even though the delay between eating and becoming ill can span many hours (Buonomano, 2011).

9 Pinker, 2014.

10 Fraps, 2014.

11 There is actually an asymmetry that works in the patron's favor. If I get exactly 21, I immediately win, and it is irrelevant if the dealer also gets 21. Of course, the probability of the cards adding to exactly 21 is below that of exceeding 21. I have also recounted this story in Buonomano, 2011.

12 Beaulieu et al., 1992; Shepherd, 1998; Herculano-Houzel, 2009.

13 I'm simplifying a bit here. There are some connections and synaptic strengths within the brain that are directly governed by our genes, but in the cortex it is likely that the strength of most synapses is determined by the interaction between synaptic learning rules and experience.

14 The first papers to describe spike-timing-dependent plasticity were Debanne et al., 1994; Markram et al., 1997; Bi and Poo, 1998, but previous work in the 1980s had hit upon similar principles (Levy and Steward, 1983). In practice, there are many different versions of the STDP rule. But in general the degree of potentiation or depression at any given interval can vary dramatically, and there is generally an asymmetry, meaning the degree of potentiation and depression at the same absolute interval is different (Abbott and Nelson, 2000; Karmarkar et al., 2002).

3:00: DAY AND NIGHT

1 Meijer and Robbers, 2014.

2 Pierce et al., 1986.

3 Routtenberg and Kuznesof, 1967; Morrow et al., 1997; Gutierrez, 2013.

4 Vitaterna et al., 1994.

5 Welsh et al., 1986; Herzog et al., 2004.

6 James, 1890. For laboratory studies on self-awakening see Moorcroft et al., 1997; Born et al., 1999; Ikeda et al., 2014.

7 http://www.nytimes.com/1989/05/17/us/isolation-researcher-loses-track-of -time-in-cave.html. *San Francisco Chronicle*. She spent 111 days alone in a cave. December 1, 1988; http://www.telegraph.co.uk/news/obituaries/science -obituaries/6216073/Maurizio-Montalbini.html.

8 Aschoff, 1985. See also Czeisler et al., 1980; Lavie, 2001.

9 Ralph et al., 1990; Weaver, 1998.

10 Johnson et al., 1998; Ouyang et al., 1998. See also Summa and Turek, 2015.

11 Nikaido and Johnson, 2000; Sharma, 2003; Rosbash, 2009. One piece of evidence that supports the view that an early force driving the evolution of circadian clocks was to optimize cell divisions to times that minimized the

deleterious effects of UV radiation is that one of the circadian light sensors in insects, *cryptochrome*, has a high degree of homology with an enzyme that repairs UV-induced DNA damage.

12 Konopka and Benzer, 1971. For an excellent popular-science account about the quest to understand the circadian clock see Reddy et al., 1984; Weiner, 1999.

13 Reddy et al., 1984. The gene was also identified at the same time by a second group (Bargiello et al., 1984).

14 In addition to the notion that temperature compensation arises from balanced temperature-dependent changes in chemical reactions (Smolen et al., 2004), it is also possible that specific amino acids within proteins lead to temperature compensation by altering their binding properties in a temperature-dependent fashion (Hussain et al., 2014).

15 While there is only one *Period* gene in *Drosophila*, in mammals there are actually three variants of the *Period* gene.

16 Colwell, 2011.

17 Davidson et al., 2006.

18 Jones et al., 1999; Toh et al., 2001; Jones et al., 2013.

19 Knutsson, 2003; Kivimäki et al., 2011.

20 Summa and Turek, 2015.

21 Sharma, 2003.

22 Aschoff, 1985.

23 There are many lines of evidence supporting the independence of circadian and second timing, including: mutations of the clock genes to not specifically affect interval timing (Cordes and Gallistel, 2008; Papachristos et al., 2011); observations in humans suggest that during prolonged circadian periods (which alter judgments of 1 hour) do not correlate with performance on second-timing tasks (Aschoff, 1985); and lesions of the SCN that dramatically alter the circadian rhythm do not alter timing on the peak interval procedure (Lewis et al., 2003). There are data that suggest the *Period* gene affects the ability of flies to properly time their courtship song (Kyriacou and Hall, 1980), but these results are controversial (Stern, 2014) and inconsistent with our understanding of how the circadian clock works. Although unlikely, it is possible that circadian-clock genes could directly contribute to other aspects of neural function that are important for timing on shorter scales. See also Golombek et al., 2014.

24 Foster and Wulff, 2005; Loh et al., 2010.

25 Foster and Roenneberg, 2008. While there is an abundance of evidence suggesting that the phase of the moon does not entrain human physiology, there are some exceptions. For example, one study suggests that the phase of the

moon entrains certain aspects of sleep physiology, such as how long it takes people to fall asleep (Cajochen et al., 2013).

26 Hoskins, 1993.

27 Tessmar-Raible et al., 2011; Zantke et al., 2013.

4:00: THE SIXTH SENSE

1 http://www.worldsciencefestival.com/2014/07/brains-twist-time-watch -deceptive-watchman/ (4/18/15).

2 Noyes and Kletti, 1972.

3 Loftus et al., 1987; Buckhout et al., 1989; Campbell and Bryant, 2007; Stetson et al., 2007; Buckley, 2014.

4 Matthews and Meck, 2016.

5 Hammond, 2012.

6 James, 1890, 624.

7 The notion that the number of events in memory determines retrospective temporal judgments is referred to as the "storage size" hypothesis (Ornstein, 1969); a related view is that it is the degree of "contextual change" during a period of time that increases retrospective estimates (Zakay and Block, 1997). Since it is contextual changes, such as alteration in the sensory stimuli, environments, or tasks, that are also likely to increase how memorable events are, these two hypotheses are highly complementary.

8 Hicks et al., 1976; Block et al., 2010.

9 Tom et al., 1997; Whiting and Donthu, 2009.

10 Van Wassenhove, 2009. This is actually a simplified account of the experiments; in actuality the standard stimuli were presented four times before the comparison or "oddball" stimulus.

11 Auditory versus visual stimuli (Wearden et al., 1998; Harrington et al., 2014). Novelty and familiarity (Tse et al., 2004; Pariyadath and Eagleman, 2007; Matthews, 2015). Intensity, size, and magnitude (Oliveri et al., 2008; Chang et al., 2011; Cai and Wang, 2014).

12 Yarrow et al., 2001; Park et al., 2003; Morrone et al., 2005.

13 James, 1890.

14 Sacks, 2004. Sacks attributes this anecdote to L. J. West (*Psychomimetic Drugs*).

15 Wearden et al., 2014; Wearden, 2015.

16 Wearden, 2015.

17 Tinklenberg et al., 1976.

18 In practice most animal studies use a variant of the fixed-interval procedure called the peak-interval procedure, in which some trials are never reinforced.

The rat study was performed by Han and Robinson, 2001. For additional studies, some confirmatory, others conflicting, see McClure and McMillan, 1997; Lieving et al., 2006; Atakan et al., 2012; Sewell et al., 2013.

19 Meck, 1996; Coull et al., 2011.

20 Rammsayer, 1992; Rammsayer and Vogel, 1992; Rammsayer, 1999; Coull et al., 2011.

21 Loftus et al., 1987; Sacks, 2004; Stetson et al., 2007; Arstila, 2012.

22 There are a number of ways this could be achieved, including: (1) depolarizing neurons by a few millivolts, so that they are closer to their action potential threshold; (2) effectively decreasing the time constant of neurons by closing the potassium leak channels; (3) increasing the amount of transmitter released from presynaptic terminals; or (4) inhibiting the inhibitory neurons, which are often dampening and slowing down the response of excitatory neurons. One might argue that it would even be possible to increase the local temperature in the brain by increasing blood flow, potentially accelerating neural processing.

23 Martin and Garfield, 2006; Terry et al., 2008; Swann et al., 2013. A further problem with the overclocking hypothesis is that even if the brain had a high-speed mode at its disposal, it is far from clear that it could be triggered quickly enough to be used in split-second life-threatening situations. For the brain to enter a hypothetical overclocking mode, sensory signals carrying news that things have just taken a terrible turn for the worse must first arrive from the sensory organs and be processed by the brain before all alarms are sounded and the brain and blood are flooded with fight-or-flight neuromodulators—including norepinephrine and epinephrine (adrenalin). Merely entering full overclocking mode could take a second.

24 Buckley, 2014.

25 Quoted from Arstila, 2012.

26 Loftus, 1996; Buonomano, 2011.

27 Cahill and McGaugh, 1996; Schacter, 1996.

28 I don't mean to downplay the importance of understanding phantom-pain syndrome, which is a very serious clinical condition, one that can cause immense suffering in amputees.

29 Arstila, 2012.

30 Wearden, 2015.

31 Noyes and Kletti, 1976.

32 A large number of papers have described the replay phenomenon, including: Wilson and McNaughton, 1994; Foster and Kokko, 2009; Karlsson and Frank, 2009.

5:00: PATTERNS IN TIME

1 Speech is highly redundant, meaning that there are generally many different cues that allow us to disambiguate ambiguous phrases. And in natural speech timing is only one of these cues; context and intonation are others. For papers that examine the role of temporal cues in speech see: Lehiste, 1960; Lehiste et al., 1976; Aasland and Baum, 2003; Schwab et al., 2008.

2 Breitenstein et al., 2001a; Breitenstein et al., 2001b; Taler et al., 2008.

3 Brownell and Gardner, 1988.

4 Aasland and Baum, 2003.

5 Grieser and Kuhl, 1988; Bryant and Barrett, 2007; Broesch and Bryant, 2015.

6 Bregman, 1990.

7 http://www.washingtonpost.com/national/jeremiah-a-denton-jr-vietnam -pow-and-us-senator-dies/2014/03/28/1a15343e-b500-11e3-b899-20667de 76985_story.html.

8 www.arrl.org/files/file/Technology/x9004008.pdf (7/8/15).

9 Wright et al., 1997. Another condition in this experiment was that subjects also improved on a "spatial" condition, that is, while they trained on a 100 ms interval bounded by two 1-kHz tones, they also improved on the discrimination of a 100 ms interval with 4-kHz tones. And subsequent studies have shown the subjects can even generalize to training on one interval to that same interval presented in a different modality—e.g., training in the somatosensory modality leads to improvement in the auditory modality (Nagarajan et al., 1998). I have not discussed these results in detail here because it appears that this generalization to different spatial channels may be dissociable from learning. Specifically, in the auditory modality generalization to different intervals only occurs after the learning of the trained intervals (Wright et al., 2010).

10 For a summary of the studies that have reported the interval discrimination learning is interval specific see Bueti and Buonomano, 2014.

11 Keele et al., 1985.

12 The 50 and 100 ms study was performed by Rammsayer et al., 2012. The drummer study was performed by Cicchini et al., 2012.

13 https://www.youtube.com/watch?v=utkb1nOJnD4 (7/14/15).

14 Patel et al., 2009.

15 Zarco et al., 2009. See also Honing et al., 2012.

16 Patel, 2006; Patel et al., 2014.

17 Meyer, 1961.

18 Doupe and Kuhl, 1999.

19 Hahnloser et al., 2002; Long et al., 2010.

20 Long and Fee, 2008.

21 Garcia and Mauk, 1998; Mauk and Buonomano, 2004; Shuler and Bear, 2006; Livesey et al., 2007; Coull et al., 2011; Bueti et al., 2012; Kim et al., 2013; Merchant et al., 2013; Crowe et al., 2014; Eichenbaum, 2014; Goel and Buonomano, 2014; Mello et al., 2015.

22 Wiener et al., 2010; Coull et al., 2011; Merchant et al., 2013; Coull et al., 2015.

23 Johnson et al., 2010; Goel and Buonomano, 2016.

24 For the sake of accuracy I should point out that the photoreceptors of the eye are not activated by light; they are actually turned off by light, as they are normally "on" when they are in the dark.

25 Chubykin et al., 2013.

26 Richards, 1973.

6:00: TIME, NEURAL DYNAMICS, AND CHAOS

1   Einstein and Infeld, 1938/1966, 180.

2   Creelman, 1962; Treisman, 1963. In the seventies and eighties more sophisticated variations of the internal clock model were developed. The most influential is referred to as *scalar expectancy theory* (SET), which in addition to a pacemaker-accumulator timekeeping mechanism includes components that store and compare temporal durations as well as a gating mechanism meant to capture the effects of attention on timing (Gibbon, 1977; Gibbon et al., 1984).

3   Feldman and Del Negro, 2006.

4   Miall, 1989; Matell and Meck, 2004; Buhusi and Meck, 2005.

5   Zucker, 1989; Zucker and Regehr, 2002.

6   Buonomano and Merzenich, 1995; Buonomano, 2000. See also Fortune and Rose, 2001.

7   Buonomano, 2000.

8   Carlson, 2009; Rose et al., 2011; Kostarakos and Hedwig, 2012. For examples of interval-selective (or, more accurately, interval-sensitive) neurons in mammals see Kilgard and Merzenich, 2002; Bray et al., 2008; Sadagopan and Wang, 2009; Zhou et al., 2010.

9   Beaulieu et al., 1992.

10 Buonomano and Merzenich, 1995; Maass et al., 2002; Buonomano and Maass, 2009.

11 Kilgard and Merzenich, 2002; Rennaker et al., 2007; Nikolić et al., 2009; Sadagopan and Wang, 2009; Zhou et al., 2010; Klampfl et al., 2012.

12 Haeusler and Maass, 2007; Buonomano and Maass, 2009; Lee and Buonomano, 2012.

13 Buonomano and Mauk, 1994; Mauk and Donegan, 1997; Medina et al., 2000).

14 Perrett et al., 1993; Raymond et al., 1996; Ohyama et al., 2003.

15 Mello et al., 2015.

16 Pastalkova et al., 2008; MacDonald et al., 2011; Kraus et al., 2013; Mac-Donald et al., 2013; Modi et al., 2014.

17 Lebedev et al., 2008; Jin et al., 2009; Crowe et al., 2010; Kim et al., 2013; Stokes et al., 2013; Crowe et al., 2014; Carnevale et al., 2015.

18 Many researchers have reported approximately linear increases in firing rate over time during timed motor tasks—tasks in which an animal makes a response after a stimulus is presented for a fixed amount of time (Quintana and Fuster, 1992; Leon and Shadlen, 2003; Mita et al., 2009; Jazayeri and Shadlen, 2015). But a study led by Michael Shadlen has suggested that these ramping patterns of activity might be best thought of as the preparation for the motor response, rather than the timer per se (although the two are often tightly correlated). For example, at the beginning of a race there may be three commands, READY – SET – GO. At the SET command, runners might start creating an expectation of when to take off, each passing moment increasing the likelihood that the GO signal will occur. Ramping neurons may encode such time-dependent expectations, which is a bit different from tracking actual time, because if animals are trained to expect a GO signal at approximately 0.15 or 1.8 seconds, the ramping cell activity goes up and down according to expectation, not according to absolute time (Janssen and Shadlen, 2005).

19 Sompolinsky et al., 1988.

20 Mante et al., 2013; Rigotti et al., 2013; Sussillo and Barak, 2013; Carnevale et al., 2015.

7:00: KEEPING TIME

1 Bhardwaj et al., 2006; Spalding et al., 2013. Studies using alternate methods were also important in demonstrating that adult neurogenesis can occur in humans (Eriksson et al., 1998).

2 In this case the half-life is given by $t_{50} = \ln(2) \times 2^{10}$ time units.

3 Duncan, 1999.

4 Matthews, 2000, 53.

5 Mumford, 1934/2010, 4.

6 In English the third line is generally sung as "morning bells are ringing," but in the original French it is "ring the morning bells." The order of the first and second lines is also different in English.

7 Matthews, 2000.

8   Matthews, 2000.

9   The standard deviations of the period of the rodent circadian clocks are esti-
    mated to be between 5 and 15 minutes (Welsh et al., 1986; Herzog et al.,
    2004.

10 Landes, 1983, 149–157.

11 Galison, 2003.

12 For example, the NIST-F2 atomic clock might lose a second every 300 mil-
    lion years. But newer-generation atomic lattice clocks can perform far better
    (Hinkley et al., 2013; Bloom et al., 2014).

13 http://www.bipm.org/en/publications/si-brochure/second.html (2/10/2015).

14 You might be wondering: if the GPS receiver needs to pick up delays of 30
    nanoseconds, doesn't it also need to have an atomic clock to compare the
    delay in the signal from the satellite? In principle, yes, but a GPS receiver can
    get away with an ordinary quartz clock by using the precise time signals from
    multiple satellites to continuously calibrate its own clock.

15 Levine, 1996, 68.

16 Mumford, 1934/2010, 14.

17 http://www.chicagotribune.com/news/chi-yellow-light-standard-change
    -20141010-story.html (2/17/2015).

18 Lombardi, 2002.

8:00: TIME: WHAT THE HELL IS IT?

1   I discuss this example and related issues in my earlier book (Buonomano,
    2011).

2   For a wonderful summary of many of the different perspectives on the nature
    of time, see Callender, 2010b.

3   Smolin, 2013.

4   Ellis, 2014.

5   Barbour, 1999, 67.

6   Muller and Nobre, 2014.

7   Callender, 2010a.

8   Penrose, 1989.

9   In order to reverse the direction of change in electromagnetism and quantum
    mechanics, other parameters also need to be flipped.

10 For an excellent presentation of the mysteries and theories relating to the
    origin of the universe and time's arrow see Carroll, 2010.

11 George Ellis is one who maintains that quantum measurement enforces an
    arrow of time (Ellis, 2008). For an excellent discussion of the measurement
    problem in quantum mechanics, and of whether these measurements are
    reversible or not, see Penrose, 1989, and Greene, 2004.

12   This is a much-shortened description of the famous double-slit experiment, which suggests that the electron does go through both slits. Specifically, even when one electron at a time is shot towards the barrier, an interference pattern is observed on the detector screen. For example, when only one slit is open, there will be some point P on the screen in which some percentage X of the electrons will hit. If we now open both slits, it would stand to reason that the same percentage X, or more, of electrons should hit that point. Oddly, an interference pattern is formed, meaning that at point P, fewer than X electrons might be detected with both slits open. Thus it seems that the electron is behaving as a wave that is interfering with itself, until the act of measurement—the collapse of the wave function. Nevertheless, if detectors had been placed at both slits, the electron would be detected at one or the other. There are many excellent popular-science books that provide great descriptions of the strange properties of the quantum world, including Rae, 1986; Greene, 2004; Carroll, 2010.

13   Even worse, the Wheeler-DeWitt equation—derived from merging quantum mechanics with general relativity—goes even further and hints at a universe in which time does not exist at all (Barbour, 1999).

9:00: THE SPATIALIZATION OF TIME IN PHYSICS

1   Zeh, 1989/2007, 199.

2   This is a thought experiment, so it is best not to dwell on the details—such as, in relation to what are we defining, the speed of the spaceship, the fact that the Earth itself is moving, and that basketball games are not generally played with open-roof stadiums.

3   More specifically, for all observers moving uniformly with respect to an inertial frame.

4   At low speeds this linear summation provides an excellent match to the true speed taking into account special relativity, which yields the speed of 399.999999999989 km/hr.

5   This equation assumes that we synchronized our watches at $t=0$, when we were in the same place; and that we define our respective positions within our own coordinate systems as $x^{you}=x^{me}=0$.

6   At this point you may be thinking to yourself: *Hang on, since the speed between both of us is the same from either perspective, the person on the train will also calculate that one year of my time will correspond to twenty-two years for the person on the train.* This issue lies at the heart of the so-called twin paradox. Comparing clocks at different points in space is an ill-advised endeavor; it is more productive to compare the clocks (or ages) after both observers are back in the same point in space—after you have returned to the platform.

Upon return, we will see that the person in the train is much younger than the person on the platform. A source of this asymmetry is that the person in the train had to change reference frames while the person on the platform remained in the same reference frame. The bottom line is that the spacetime interval traveled by the person on the platform is *more* than the spacetime interval traveled by the person on the train—and this spacetime interval corresponds to clock time (so-called proper time). For a discussions of the twin paradox see Lockwood, 2005; Lasky, 2012.

7　Hafele and Keating, 1972b, a. The earth is of course not a static body: for example, it is moving in relation to the sun. But for our purposes we can consider the center of earth a valid fixed frame of reference. But because the earth is rotating, the speed of the eastbound plane should be added to the rotational speed of the earth; thus eastbound clocks were going faster than their landbound cousins, resulting in a slowing of the clocks. Hafele and Keating also put clocks on westbound flights, and since westbound flights effectively counteracted the rotational speed of the earth, these clocks exhibited the expected kinematic speedup produced by special relativity. Because the clocks in the planes also were under weaker gravitational fields, they also had to take into account the predictions of general relativity, which were also confirmed in this study.

8　Because special relativity states that time dilates and space contracts at high speeds, the length of the train will actually be longer from my perspective. I have simplified this thought experiment by ignoring this fact. But space contraction does not alter the results, because from my frame of reference you will still be standing in the middle of the train, and the backward and forward bullets still start from the same distances from the back and front of the train, respectively.

9　In this example I've placed both observers the same distance from the back and front of the train, in an attempt to convey that the transmission delays from the front and back should be the same. But we can imagine having an array of synchronized clocks arranged all along the platform that will be stopped by a nearby window breaking. These clocks will tell us the front and back window broke at different times.

10　For a more technical and historical discussion of the relativity of simultaneity see Brown, 2005.

11　Rietdijk, 1966; Putnam, 1967.

12　It is important to note that the loss of absolute simultaneity is an argument that certainly favors the notion of a block universe—but that is all it is. One might argue that the real lesson to be learned from special relativity is not that we live in a block universe, but that the concept of simultaneity is a vestige from Newton's notion of absolute time. Perhaps it does not really make sense to ask whether two distant events are simultaneous: after all, the only way to determine if two distant events are simultaneous or not is with clocks, and clocks are simply devices that measure change in a local volume of space.

13 Karl Popper also recounts a discussion in which Einstein confirmed that he accepted the block universe view (Popper, 1992), and that in the conversation he referred to Einstein as Parmenides.

14 Cited in Prigogine and Stengers, 1984, 214.

15 Penrose, 1989, 394.

16 Davies, 1995, 283.

17 Barbour, 1999, 267.

18 Greene, 2004.

19 Schuster et al., 2004.

20 Panagiotaropoulos et al., 2012; Kandel et al., 2013; Purdon et al., 2013; Baker et al., 2014; Ishizawa et al., 2016.

21 Many physical processes, including life, temperature, or the velocity of a particle, are defined by how a system changes over time. But it is important to understand that none of these represent arguments against eternalism because eternalism accepts that change is happening over the temporal axis of spacetime! I'm suggesting that consciousness is fundamentally different because according to the moments-within-a-moment hypothesis of Barbour and Greene, we must be conscious within a single "frame."

22 Pinker's full quote: "It's almost impossible to imagine abolishing time from one's awareness, leaving the last though immobilized like a stuck car horn, while continuing to have a mind at all. For Descartes the distinction between the physical and the mental depended on this difference. Matter is extended in space, but consciousness exists in time as surely as it proceeds from 'I think' to 'I am'" (Pinker, 2007).

23 Lockwood, 2005.

24 The equation of spatial relativity reveals that if hypothetical particles that traveled faster than the speed of light (tachions) existed, it would be possible to send signals back in time—potentially altering the past. Strictly speaking this would not be a form of time *travel*, but communication with the past and future.

25 Greene, 2004.

10:00: THE SPATIALIZATION OF TIME IN NEUROSCIENCE

1 Quoted from Papert, 1999. I was not able to find independent confirmation of this quote.

2 Quoted from Droit-Volet, 2003.

3 Piaget, 1946/1969.

4 Siegler and Richards, 1979. See also Matsuda, 1996.

5 Piaget, 1946/1969, 279.

6 Walsh, 2003; Núñez and Cooperrider, 2013; Bender and Beller, 2014.

7 Núñez and Cooperrider, 2013.

8 Lakoff and Johnson, 1980/2003.

9 Núñez and Sweetser, 2006.

10 McGlone and Harding, 1998; Boroditsky and Ramscar, 2002.

11 Lakoff and Johnson, 1980/2003.

12 Price-Williams, 1954.

13 Huang and Jones, 1982. Kappa and tau effects have been demonstrated in many different studies in both the visual and somatosensory modalities (Helson and King, 1931; Cohen et al., 1953; Sarrazin et al., 2004; Goldreich, 2007; Grondin et al., 2011).

14 Casasanto and Boroditsky, 2008. For a similar study that also established an asymmetric relationship between how distance influences temporal judgments, and vice versa, see Coull et al., 2015.

15 Walsh, 2003; Bueti and Walsh, 2009.

16 Xuan et al., 2007; Hayashi et al., 2013; Cai and Wang, 2014.

17 Ishihara et al., 2008; Kiesel and Vierck, 2009.

18 Saj et al., 2014.

19 Pastalkova et al., 2008; Kraus et al., 2013; Genovesio and Tsujimoto, 2014.

20 See Calaprice, 2005.

21 It has been pointed out that the kappa effect holds parallels to special relativity in the sense that if we consider the subject in a kappa experiment to be observing a fast-moving object in a different reference frame, her clock will effectively be going fast—that is, she will have measured more time to have elapsed, similar to the stay-at-rest twin in the twin paradox (Goldreich, 2007). But on the other hand, in special relativity there is an absolute trade-off between distance and time. Locally, speed and elapsed time are inversely related to each other, whereas in the kappa effect perceived duration and speed are proportional to each other.

22 For example, Piaget stated: "As paradox as it may seem, the relative durations and the proper times of Einstein's theory relate to absolute time as absolute time to the individual times and local times of the child's intuition." Quoted from Sauer, 2014. And "In the macroscopic universe, however, the subordination of time with respect to velocity remains fundamental since at high velocities relativistic time comes up against the same difficulties as does the young child's idea of time, and also presupposes a subordination of temporal relationships with respect to certain velocities." (Piaget, 1972).

23 Please note that this is an entirely different point from the one I made in the

previous chapter relating to the subjective sense of the passage of time. Independent of the distortions imposed by a myriad of temporal illusions, our subjective sense of the passage of time does need to be explained in a manner compatible with both physics and neuroscience.

24 The general strategy of using prior experiences together with current best estimates is referred to as Bayesian decision theory (Kording, 2007), and is believed to account for many aspects of perception and decision making, including making temporal judgments (Collyer, 1976; Goldreich, 2007; Jazayeri and Shadlen, 2010).

25 Smolin, 2013.

26 This point is subtly different from the fact that intuitively we favor presentism because only the present seems to be real. Here the point is that given the abstract and mathematical representations of time as a dimension much like space, perhaps we are biased toward eternalism because humans seem to conceptualize time in terms of space.

11:00: MENTAL TIME TRAVEL

1 http://www.nytimes.com/2011/04/21/world/asia/21stones.html (5/15/2015).

2 Suddendorf and Corballis, 1997, 2007.

3 Tulving, 2005.

4 Taylor et al., 1994.

5 Hassabis et al., 2007; Race et al., 2011; Kwan et al., 2012.

6 Tulving, 1985.

7 Gilbert, 2007.

8 Clayton and Dickinson, 1999; Raby et al., 2007; Clayton et al., 2009.

9 Osvath and Persson, 2013; Bourjade et al., 2014; Scarf et al., 2014. For two popular books that discuss the issue of mental time travel in animals see Corballis, 2011; Suddendorf, 2013.

10 http://www.spectator.co.uk/features/5896113/if-we-have-souls-then-so-do -chimps/ (5/15/2015).

11 Gordon, 2004. "The Pirahãs have no notion of their age, nor of time concepts like 'how long have you known. . . .'" Personal communication from Daniel Everett (3/4/2009).

12 Everett, 2008, 132.

13 Colapinto, 2007.

14 Psychologists have attempted to create a taxonomy of people's temporal outlooks. The Zimbardo Time Perspective Inventory, for example, asks people to indicate on a five-point scale the degree to which they agree with statements of the sort: *You can't really plan for the future because things change so much*;

It gives me pleasure to think of my past; Things rarely work out as I expected. Depending on the answers people are assigned past-negative, past-positive, present-fatalistic, present-hedonistic, future perspectives (Zimbardo and Boyd, 2008).

15 http://www.cbsnews.com/news/sea-gypsies-saw-signs-in-the-waves/ (5/15/2015).

16 http://money.usnews.com/money/blogs/the-best-life/2013/06/20/retirement -shortfall-may-top-14-trillion (12/9/2015).

17 James Surowiecki provides a brief discussion of the chronic problem of pension funds in Surowiecki, 2013. The quote from Mark Twain is also taken from his piece.

18 Kable and Glimcher, 2007.

19 Critchfield and Kollins, 2001; Wittmann and Paulus, 2007; Seeyave et al., 2009; MacKillop et al., 2011.

20 Frederick et al., 2002.

21 Prelec and Simester, 2001; Raghubir and Srivastava, 2008.

22 I say this because ultimately such rewards are paid from the credit-card fees retailers pay, and retailers must set their prices taking these fees into account.

23 Buonomano, 2011, ch. 4.

24 Peters and Büchel, 2010. See also Hakimi and Hare, 2015.

25 Herculano-Houzel, 2009; Fox, 2011.

26 Purves et al., 2008.

27 Jacobs et al., 2001; Wood and Grafman, 2003; Wise, 2008; Fuster and Bressler, 2014.

28 Atance and O'Neill, 2001; Fuster and Bressler, 2014.

29 Sellitto et al., 2010; Peters, 2011). I have oversimplified here a bit as the prefrontal cortex is actually further subdivided into a number of distinct regions, some of which have been postulated to be partial toward short-term rewards.

30 McClure et al., 2004.

31 Botzung et al., 2008; Benoit and Schacter, 2015.

32 Hassabis et al., 2007; Race et al., 2011; Kwan et al., 2012.

33 Gilbert, 2007; Killingsworth and Gilbert, 2010.

34 Everett, 2008, 273.

12:00: CONSCIOUSNESS: BINDING THE PAST AND THE FUTURE

1 Burr et al., 1994; Yarrow et al., 2001.

2 Koch, 2004.

3 Kanabus et al., 2002; Alais and Cass, 2010.

4 Van Wassenhove et al., 2007; Mégevand et al., 2013. There is a further factor

that I will not elaborate on here relating to the delays it takes the ear and eye to process auditory and visual signals. Vision is actually fairly slow, compared to audition.

5   I have oversimplified this story a bit; in reality the visual system does, in effect, have a built-in delay. Visual information from the retina may arrive in the visual cortex a full 50 ms after information from the cochlea arrives in the auditory cortex. This hardwired delay is mostly a product of the fact that the phototransduction of the retina is much slower than the mechanical transduction of the cochlea.

6   Fujisaki et al., 2004; Toida et al., 2014; Van der Burg et al., 2015.

7   Geldard and Sherrick, 1972; Kilgard and Merzenich, 1995; Goldreich and Tong, 2013.

8   Dennett, 1991; Buonomano, 2011; Herzog et al., 2016.

9   See for example Dehaene and Changeux, 2011; Kandel, 2013.

10   Lamy et al., 2009). See also Salti et al., 2015.

11   Dehaene, 2014, 126. For another example of how late manipulations, 400 ms after a stimulus, can alter the conscious perception of the stimuli, see Scharnowski et al., 2009; Sergent et al., 2013.

12   There is a long and venerable philosophical history on the topic of free will, and more recently on the neuroscience of free will. As an introduction I recommend the following articles: Montague, 2008; Haggard, 2011; Nichols, 2011; Smith, 2011; and books: Dennett, 2003; Harris, 2012.

13   Definition #2, http://www.oed.com/view/Entry/74438 (12/30/2015).

14   Montague, 2008.

15   Hawking, 1996.

16   Penrose, 1989, 558.

17   Lockwood, 2005.

18   Although it has been argued that this feeling of choice is an illusion that collapses upon closer analysis (Harris, 2012).

19   Wegner, 2002.

20   Hume, 1739/2000.

21   Huxley, 1894/1911, 244.

22   Fried et al., 2011.

23   Libet et al., 1983; Lau et al., 2007; Haggard, 2008; Soon et al., 2008; Murakami et al., 2014.

24   Haggard, 2011.

25   Dehaene, 2014, 91. The author Adam Gopnik has expressed this notion using the spokesperson metaphor: "What we call consciousness is just an illusion, and bears the same relation to the working of our real minds that the White House press spokesman bears to the workings of the Bush White House: it is

there to find rationalization and systematic reasons for feelings and decisions made by dim, hidden powers of whose pettish and irrational purposes it is aware of only long after the fact" (*The New Yorker*, July 4, 2005).

26 Gazzaniga and Steven, 2005; Gazzaniga, 2011.

27 Quoted from "Tomorrow never was" by Zeeya Merali (*Discover*, June 2015).

28 Nichols, 2011; Shariff and Vohs, 2014.

29 Within the context of the Model Penal Code these three scenarios would roughly correspond to the following mental states: purposely, recklessly/negligently, and strict liability.

30 Here I'm referring to the Wheeler-DeWitt equation, which is aimed at merging general relativity and quantum mechanics. To the bewilderment of many, this attempt resulted in an equation in which there is no time parameter, apparently leading to the conclusion that time itself does not exist—a view advanced by physicists such as Julian Barbour and Carlos Rovelli. For excellent presentations of time in quantum mechanics and the Wheeler-DeWitt equation see Barbour, 1999; Rovelli, 2004; Lockwood, 2005; Callender, 2010a; Smolin, 2013.

31 Koch, 2004; Dehaene, 2014.

BIBLIOGRAPHY

Aasland, W. A., Baum, S. R. (2003). Temporal parameters as cues to phrasal boundaries: A comparison of processing by left- and right-hemisphere brain-damaged individuals. *Brain and Language, 87,* 385–399.

Abbott, L. F., Nelson, S. B. (2000). Synaptic plasticity: Taming the beast. *Nature Neuroscience, 3,* 1178–1183.

Alais, D., Cass, J. (2010). Multisensory perceptual learning of temporal order: Audiovisual learning transfers to vision but not audition. *PLoS ONE, 5,* e11283.

Arstila, V. (2012). Time slows down during accidents. *Frontiers in Psychology, 3,* 196.

Aschoff, J. (1985). On the perception of time during prolonged temporal isolation. *Human Neurobiology, 4,* 41–52.

Atakan, Z., Morrison, P., Bossong, M. G., Martin-Santos, R., Crippa, J. A. (2012). The effect of cannabis on perception of time: A critical review. *Current Pharmaceutical Design, 18,* 4915–4922.

Atance, C. M., O'Neill, D. K. (2001). Episodic future thinking. *Trends in Cognitive Sciences, 5,* 533–539.

Baker, R., Gent, T. C., Yang, Q., Parker, S., Vyssotski, A. L., Wisden, W., Brickley, S. G., Franks, N. P. (2014). Altered activity in the central medial thalamus precedes changes in the neocortex during transitions into both sleep and propofol anesthesia. *Journal of Neuroscience, 34,* 13326–13335.

Barbour, J. (1999). *The end of time: The next revolution in physics.* New York: Oxford University Press.

Bargiello, T. A., Jackson, F. R., Young, M. W. (1984). Restoration of circadian behavioural rhythms by gene transfer in Drosophila. *Nature, 312,* 752–754.

Beaulieu, C., Kisvarday, Z., Somogyi, P., Cynader, M., Cowey, A. (1992). Quantitative distribution of GABA-immunopositive and -immunonegative neurons and synapses in the monkey striate cortex (area 17). *Cerebral Cortex, 2,* 295–309.

Bender, A., Beller, S. (2014). Mapping spatial frames of reference onto time: A review of theoretical accounts and empirical findings. *Cognition, 132*, 342–382.

Benoit, R. G., Schacter, D. L. (2015). Specifying the core network supporting episodic simulation and episodic memory by activation likelihood estimation. *Neuropsychologia, 75*, 450–457.

Bhardwaj, R. D., Curtis, M. A., Spalding, K. L., Buchholz, B. A., Fink, D., Björk-Eriksson, T., Nordborg, C., Gage, F. H., Druid, H., Eriksson, P. S., Frisén, J. (2006). Neocortical neurogenesis in humans is restricted to development. *Proceedings of the National Academy of Sciences, 103*, 12564–12568.

Bi, G. Q., Poo, M. M. (1998). Synaptic modifications in cultured hippocampal neurons: dependence on spike timing, synaptic strength, and postsynaptic cell type. *Journal of Neuroscience, 18*, 10464–10472.

Block, R. A., Hancock, P. A., Zakay, D. (2010). How cognitive load affects duration judgments: A meta-analytic review. *Acta Psychologica, 134*, 330–343.

Bloom, B. J., Nicholson, T. L., Williams, J. R., Campbell, S. L., Bishof, M., Zhang, X., Zhang, W., Bromley, S. L., Ye, J. (2014). An optical lattice clock with accuracy and stability at the 10-18 level. *Nature*, advance online publication.

Born, J., Hansen, K., Marshall, L., Molle, M., Fehm, H. L. (1999). Timing the end of nocturnal sleep. *Nature, 397*, 29–30.

Boroditsky, L., Ramscar, M. (2002). The roles of body and mind in abstract thought. *Psychological Science, 13*, 185–189.

Botzung, A., Denkova, E., Manning, L. (2008). Experiencing past and future personal events: Functional neuroimaging evidence on the neural bases of mental time travel. *Brain and Cognition, 66*, 202–212.

Bourjade, M., Call, J., Pele, M., Maumy, M., Dufour, V. (2014). Bonobos and orangutans, but not chimpanzees, flexibly plan for the future in a token-exchange task. *Animal Cognition, 17*, 1329–1340.

Bray, S., Rangel, A., Shimojo, S., Balleine, B., O'Doherty, J. P. (2008). The neural mechanisms underlying the influence of pavlovian cues on human decision making. *Journal of Neuroscience, 28*, 5861–5866.

Bregman, A. S. (1990). *Auditory scene analysis: The perceptual organization of sound*. Cambridge: MIT Press.

Breitenstein, C., Van Lancker, D., Daum, I. (2001a). The contribution of speech rate and pitch variation to the perception of vocal emotions in a German and an American sample. *Cognition and Emotion, 15*, 57–79.

Breitenstein, C., Van Lancker, D., Daum, I., Waters, C. H. (2001b). Impaired perception of vocal emotions in Parkinson's disease: influence of speech time processing and executive functioning. *Brain and Cognition, 45*, 277–314.

Brown, H. (2005). *Physical relativity: space-time structure from a dynamical perspective*. Oxford: Oxford University Press.

Brownell, H. H., Gardner, H. (1988). Neuropsychological insights into humour. In: *Laughing matters: A serious look at humour.* (Durant, J., Miller, J., eds). Essex: Longman Scientific & Technical.

Buckhout, R., Fox, P., Rabinowitz, M. (1989). Estimating the duration of an earthquake: Some shaky field observations. *Bulletin of the Psychonomic Society, 27,* 375–378.

Buckley, R. (2014). Slow time perception can be learned. *Frontiers in Psychology, 5.*

Bueti, D., Walsh, V. (2009). The parietal cortex and the representation of time, space, number and other magnitudes. *Philosophical Transactions of the Royal Society of London B: Biological Sciences, 364,* 1831–1840.

Bueti, D., Buonomano, D. V. (2014). Temporal perceptual learning. *Timing and Time Perception, 2,* 261–289.

Bueti, D., Lasaponara, S., Cercignani, M., Macaluso, E. (2012). Learning about time: Plastic changes and interindividual brain differences. *Neuron, 75,* 725–737.

Buhusi, C. V., Meck, W. H. (2005). What makes us tick? Functional and neural mechanisms of interval timing. *Nature Reviews Neuroscience, 6,* 755–765.

Buonomano, D. V. (2000). Decoding temporal information: a model based on short-term synaptic plasticity. *Journal of Neuroscience, 20,* 1129–1141.

———. (2011). *Brain bugs: How the brain's flaws shape our lives.* New York: W. W. Norton.

Buonomano, D. V., Mauk, M. D. (1994). Neural network model of the cerebellum: temporal discrimination and the timing of motor responses. *Neural Computation, 6,* 38–55.

Buonomano, D. V., Merzenic, M. M. (1995). Temporal information transformed into a spatial code by a neural network with realistic properties. *Science, 267,* 1028–1030.

Buonomano, D. V., Maass, W. (2009). State-dependent computations: Spatiotemporal processing in cortical networks. *Nature Reviews Neuroscience, 10,* 113–125.

Burr, D. C., Morrone, M. C., Ross, J. (1994). Selective suppression of the magnocellular visual pathway during saccadic eye movements. *Nature, 371,* 511–513.

Cahill, L., McGaugh, J. L. (1996). Modulation of memory storage. *Current Opinion in Neurobiology, 6,* 237–242.

Cai, Z. G., Wang, R. (2014). Numerical magnitude affects temporal memories but not time encoding. *PLoS ONE, 9,* e83159.

Cajochen, C., Altanay-Ekici, S., Münch, M., Frey, S., Knoblauch, V., Wirz-Justice, A. (2013). Evidence that the lunar cycle influences human sleep. *Current Biology, 23,* 1485–1488.

Calaprice, A. (2005). *The new quotable Einstein.* Princeton: Princeton University Press.

Callender, C. (2010a). Is time an illusion? *Scientific American, June*, 59–65.

Callender, C., Edney, R. (2010b). *Introducing time: A graphic guide*. London: Icon Books.

Campbell, L. A., Bryant, R. A. (2007). How time flies: A study of novice skydivers. *Behaviour Research and Therapy, 45*, 1389–1392.

Carlson, B. A. (2009). Temporal-pattern recognition by single neurons in a sensory pathway devoted to social communication behavior. *Journal of Neuroscience, 29*, 9417–9428.

Carnevale, F., de Lafuente, V., Romo, R., Barak, O., Parga, N. (2015). Dynamic control of response criterion in premotor cortex during perceptual detection under temporal uncertainty. *Neuron, 86*, 1067–1077.

Carroll, S. (2010). *From eternity to here: The quest for the ultimate theory of time*. New York: Penguin.

Chang, A. Y.-C., Tzeng, O. J. L., Hung, D. L., Wu, D. H. (2011). Big time is not always long: Numerical magnitude automatically affects time reproduction. *Psychological Science, 22*, 1567–1573.

Chubykin, Alexander A., Roach, Emma B., Bear, Mark F., Shuler, Marshall G. H. (2013). A cholinergic mechanism for reward timing within primary visual cortex. *Neuron, 77*, 723–735.

Cicchini, G. M., Arrighi, R., Cecchetti, L., Giusti, M., Burr, D. C. (2012). Optimal encoding of interval timing in expert percussionists. *Journal of Neuroscience, 32*, 1056–1060.

Clark, A. (2013). Whatever next? Predictive brains, situated agents, and the future of cognitive science. *Behavioral and Brain Sciences, 36*, 181–204.

Clayton, N. S., Dickinson, A. (1999). Scrub jays (Aphelocoma coerulescens) remember the relative time of caching as well as the location and content of their caches. *Journal of Comparative Psychology, 113*, 403–416.

Clayton, N. S., Russell, J., Dickinson, A. (2009). Are animals stuck in time or are they chronesthetic creatures? *Topics in Cognitive Science, 1*, 59–71.

Cohen, J., Hansel, C. E., Sylvester, J. D. (1953). A new phenomenon in time judgment. *Nature, 172*, 901.

Colapinto, J. (2007). The interpreter. *The New Yorker*, April 16, 118–137.

Collyer, C. E. (1976). The induced asynchrony effect: Its role in visual judgments of temporal order and its relation to other dynamic perceptual phenomena. *Perception & Psychophysics, 19*, 47–54.

Colwell, C. S. (2011). Linking neural activity and molecular oscillations in the SCN. *Nature Reviews Neuroscience, 12*, 553–569.

Corballis, M. C. (2011). *The recursive mind: the origins of human language thought and civilization*. Princeton, NJ: Princeton University Press.

Cordes, S., Gallistel, C. R. (2008). Intact interval timing in circadian CLOCK mutants. *Brain Research, 1227*, 120–127.

Coull, J. T., Cheng, R.-K., Meck, W. H. (2011). Neuroanatomical and neuro-chemical substrates of timing. *Neuropsychopharmacology, 36*, 3–25.

Coull, J. T., Charras, P., Donadieu, M., Droit-Volet, S., Vidal, F. (2015). SMA selectively codes the active accumulation of temporal, not spatial, magnitude. *Journal of Cognitive Neuroscience, 27*, 2281–2298.

Creelman, C. D. (1962). Human discrimination of auditory duration. *Journal of the Acoustical Society of America, 34*, 582–593.

Critchfield, T. S., Kollins, S. H. (2001). Temporal discounting: basic research and the analysis of socially important behavior. *Journal of Applied Behavior Analysis, 34*, 101–122.

Crowe, D. A., Averbeck, B. B., Chafee, M. V. (2010). Rapid sequences of popula-tion activity patterns dynamically encode task-critical spatial information in parietal cortex. *Journal of Neuroscience, 30*, 11640–11653.

Crowe, D. A., Zarco, W., Bartolo, R., Merchant, H. (2014). Dynamic representa-tion of the temporal and sequential structure of rhythmic movements in the primate medial premotor cortex. *Journal of Neuroscience, 34*, 11972–11983.

Czeisler, C., Weitzman, E., Moore-Ede, M., Zimmerman, J., Knauer, R. (1980). Human sleep: its duration and organization depend on its circadian phase. *Science, 210*, 1264–1267.

Davidson, A. J., Sellix, M. T., Daniel, J., Yamazaki, S., Menaker, M., Block, G. D. (2006). Chronic jet-lag increases mortality in aged mice. *Current Biol-ogy, 16*, R914–916.

Davies, P. (1995). *About time: Einstein's unfinished revolution.* New York: Simon & Schuster.

———. (2012). That mysterious flow. *Scientific American, 21*, 8–13.

Debanne, D., Gahwiler, B. H., Thompson, S. M. (1994). Asynchronous pre- and postsynaptic activity induces associative long-term depression in area CA1 of the rat hippocampus in vitro. *Proceedings of the National Academy of Science USA, 91*, 1148–1152.

Dehaene, S. (2014). *Consciousness and the brain: Deciphering how the brain codes our thoughts.* New York: Viking.

Dehaene, S., Changeux, J.-P. (2011). Experimental and theoretical approaches to conscious processing. *Neuron, 70*, 200–227.

Dennett, D. C. (1991). *Consciousness explained.* New York: Little, Brown and Company.

———. (2003). *Freedom evolves.* New York: Penguin Books.

DiCarlo, J. J., Cox, D. D. (2007). Untangling invariant object recognition. *Trends in Cognitive Sciences, 11*, 333–341.

Doupe, A. J., Kuhl, P. K. (1999). Birdsong and human speech: common themes and mechanisms. *Annual Review of Neuroscience, 22*, 567–631.

Droit-Volet, S. (2003). Temporal experience and timing in children. In: *Func-*

tional and neural mechanisms of interval timing (Meck, WH, ed.), 183–288. Boca Raton: CRC Press.

Dudai, Y., Carruthers, M. (2005). The Janus face of Mnemosyne. *Nature, 434,* 567.

Duncan, D. E. (1999). *Calendar: humanity's epic struggle to determine a true and accurate year.* New York: Avon Books.

Eichenbaum, H. (2014). Time cells in the hippocampus: a new dimension for mapping memories. *National Review of Neuroscience, 15,* 732–744.

Einstein, A. (1905). On the electrodynamics of moving bodies. *Annalen der Physik, 17,* 891–921.

Einstein, A., Infeld, L. (1938/1966). *The evolution of physics.* New York: Simon & Schuster.

Ellis, G. F. R. (2008). On the flow of time. *FXQi Essay.* http://fqxiorg/data/essay -contest-files/Ellis_Fqxi_essay_contest__Epdf.

———. (2014). The evolving block universe and the meshing together of times. *Annals of the New York Academy of Sciences, 1326,* 26–41.

Eriksson, P. S., Perfilieva, E., Bjork-Eriksson, T., Alborn, A. M., Nordborg, C., Peterson, D. A., Gage, F. H. (1998). Neurogenesis in the adult human hippocampus. *Nature Medicine, 4,* 1313–1317.

Everett, D. (2008). *Don't sleep, there are snakes.* New York: Pantheon.

Feldman, J. L., Del Negro, C. A. (2006). Looking for inspiration: new perspectives on respiratory rhythm. *National Review of Neuroscience, 7,* 232–241.

Feldmeyer, D., Lubke, J., Silver, R. A., Sakmann, B. (2002). Synaptic connections between layer 4 spiny neurone-layer 2/3 pyramidal cell pairs in juvenile rat barrel cortex: physiology and anatomy of interlaminar signalling within a cortical column. *Journal of Physiology, 538,* 803–822.

Földiák, P. (1991). Learning invariance from transformation sequences. *Neural Computation, 3,* 194–200.

Fortune, E. S., Rose, G. J. (2001). Short-term synaptic plasticity as a temporal filter. *Trends in Neurosciences, 24,* 381–385.

Foster, K. R., Kokko, H. (2009). The evolution of superstitious and superstition-like behaviour. *Proceedings of the Royal Society B: Biological Sciences, 276,* 31–37.

Foster, R. G., Wulff, K. (2005). The rhythm of rest and excess. *National Review of Neuroscience, 6,* 407–414.

Foster, R. G., Roenneberg, T. (2008). Human responses to the geophysical daily, annual and lunar cycles. *Current Biology, 18,* R784–R794.

Fox, D. (2011). The limits of intelligence. *Scientific American,* July, 36-43.

Fraisse, P. (1963). *The psychology of time.* New York: Harper & Row.

Fraps, T. (2014). Time and magic-manipulating subjective temporality. In: *Subjective time: the philosophy, psychology, and neuroscience of temporality* (Arstila, V., Lloyd, D., eds.), 263–285. Cambridge, MA: MIT Press.

Frederick, S., Loewenstein, G., O'Donoghue, T. (2002). Time discounting and time preference: a critical review. *Journal of Economic Literature*, *45*, 351–401.

Fried, I., Mukamel, R., Kreiman G. (2011). Internally generated preactivation of single neurons in human medial frontal cortex predicts volition. *Neuron*, *69*, 548–562.

Fujisaki, W., Shimojo, S., Kashino, M., Nishida S. (2004). Recalibration of audio-visual simultaneity. *Nature Neuroscience*, *7*, 773–778.

Fuster, J. M., Bressler, S. L. (2014). Past makes future: Role of pFC in prediction. *Journal of Cognitive Neuroscience*, *27*, 639–654.

Galison, P. (2003). *Einstein's clocks and Poincaré's maps: Empires of time*. New York: W. W. Norton.

Garcia, K. S., Mauk, M. D. (1998). Pharmacological analysis of cerebellar contributions to the timing and expression of conditioned eyelid responses. *Neuropharmacology*, *37*, 471–480.

Gazzaniga, M. S. (2011). Neuroscience in the courtroom. *Scientific American*, April, 54–59.

Gazzaniga, M. S., Steven, M. S. (2005). Neuroscience and the law. *Scientific American Mind*, *16*, 42–49.

Geldard, F. A., Sherrick, C. E. (1972). The cutaneous "rabbit": A perceptual illusion. *Science*, *178*, 178–179.

Genovesio, A., Tsujimoto, S. (2014). From duration and distance comparisons to goal encoding in prefrontal cortex. In: *Neurobiology of Interval Timing* (Merchant, H., de Lafuente, V., eds.), 167–186. New York: Springer.

Gibbon, J. (1977). Scalar expectancy theory and Weber's law in animal timing. *Psychological Review*, *84*, 279–325.

Gibbon, J., Church, R. M., Meck, W. H. (1984). Scalar timing in memory. *Annals of the New York Academy of Science*, *423*, 52–77.

Gilbert, D. (2007). *Stumbling on happiness*. New York: Vintage Books.

Goel, A., Buonomano, D. V. (2014). Timing as an intrinsic property of neural networks: evidence from in vivo and in vitro experiments. *Philosophical Transactions of the Royal Society of London B: Biological Science*, *369*, 20120460.

———. (2016). Temporal interval learning in cortical cultures is encoded in intrinsic network dynamics. *Neuron*, *91*, 320–327.

Goldreich, D. (2007). A Bayesian perceptual model replicates the cutaneous rabbit and other tactile spatiotemporal illusions. *PLoS ONE*, *2*, e333.

Goldreich, D., Tong, J. (2013). Prediction, postdiction, and perceptual length contraction: a Bayesian low-speed prior captures the cutaneous rabbit and related illusions. *Frontiers in Psychology*, *4*, 579.

Golombek, D. A., Bussi, I. L., Agostino, P. V. (2014). Minutes, days and years: molecular interactions among different scales of biological timing. *Philosophical Transactions of the Royal Society B: Biological Sciences*, *369*, 20120465.

Gordon, P. (2004). Numerical cognition without words: evidence from Amazonia. *Science, 306*, 496–499.

Greene, B. (2004). *The fabric of the cosmos: Space, time, and the texture of reality.* New York: Vintage Books.

Grondin, S., Kuroda, T., Mitsudo, T. (2011). Spatial effects on tactile duration categorization. *Canadian Journal of Experimental Psychology/Revue canadienne de psychologie expérimentale, 65*, 163–167.

Gutierrez, E. (2013). A rat in the labyrinth of anorexia nervosa: Contributions of the activity-based anorexia rodent model to the understanding of anorexia nervosa. *International Journal of Eating Disorders, 46*, 289–301.

Haeusler, S., Maass, W. (2007). A statistical analysis of information-processing properties of lamina-specific cortical microcircuit models. *Cerebral Cortex, 17*, 149–162.

Hafele, J. C., Keating, R. E. (1972a). Around-the-world atomic clocks: Observed relativistic time gains. *Science, 177*, 168–170.

———. (1972b). Around-the-world atomic clocks: Predicted relativistic time gains. *Science, 177*, 166–168.

Haggard, P. (2008). Human volition: towards a neuroscience of will. *National Review of Neuroscience, 9*, 934–946.

———. (2011). Decision time for free will. *Neuron, 69*, 404–406.

Hahnloser, R. H. R., Kozhevnikov, A. A., Fee M. S. (2002). An ultra-sparse code underlies the generation of neural sequence in a songbird. *Nature, 419*, 65–70.

Hakimi, S., Hare, T. A. (2015). Enhanced neural responses to imagined primary rewards predict reduced monetary temporal discounting. *Journal of Neuroscience, 35*, 13103–13109.

Hammond, C. (2012). *Time warped: Unlocking the mysteries of time perception.* New York: Harper-Perennial.

Han, C. J., Robinson, J. K. (2001). Cannabinoid modulation of time estimation in the rat. *Behavioral Neuroscience 115*, 243–246.

Harrington, D. L., Castillo, G. N., Reed J. D., Song, D. D., Litvan I., Lee, R. R. (2014). Dissociation of neural mechanisms for intersensory timing deficits in Parkinson's disease. *Timing & Time Perception, 2*, 145–168.

Harris, S. (2012). *Free will.* New York: Free Press.

Hassabis, D., Kumaran, D., Vann, S. D., Maguire, E. A. (2007). Patients with hippocampal amnesia cannot imagine new experiences. *Proceedings of the National Academy of Sciences USA, 104*, 1726–1731.

Hawking, S. (1996). *A brief history of time.* New York: Bantam Books.

Hayashi, M. J., Kanai, R., Tanabe, H. C., Yoshida, Y., Carlson, S., Walsh, V., Sadato, N. (2013). Interaction of numerosity and time in prefrontal and parietal cortex. *Journal of Neuroscience, 33*, 883–893.

Helson, H., King, S. M. (1931). The tau effect: an example of psychological relativity. *Journal of Experimental Psychology, 14*, 202–217.

Henderson, J., Hurly, T. A., Bateson, M., Healy, S. D. (2006). Timing in free-living rufous hummingbirds, Selasphorus rufus. *Current Biology, 16*, 512–515.

Herculano-Houzel, S. (2009). The human brain in numbers: a linearly scaled-up primate brain. *Frontiers in Human Neuroscience, 3.*

Herzog, E. D., Aton, S. J., Numano, R., Sakaki, Y., Tei H. (2004). Temporal precision in the mammalian circadian system: A reliable clock from less reliable neurons. *Journal of Biological Rhythms, 19*, 35–46.

Herzog M. H., Kammer T., Scharnowski F. (2016). Time slices: What is the duration of a percept? *PLoS Biol, 14*, e1002433.

Hicks, R. E., Miller, G. W., Kinsbourne, M. (1976). Prospective and retrospective judgments of time as a function of amount of information processed. *American Journal of Psychology, 89*, 719–730.

Hinkley, N., Sherman, J. A., Phillips, N. B., Schioppo, M., Lemke, N. D., Beloy, K., Pizzocaro, M., Oates, C. W., Ludlow, A. D. (2013). An atomic clock with 10–18 instability. *Science, 341*, 1215–1218.

Honing, H., Merchant, H., Háden, G. P., Prado, L., Bartolo, R. (2012). Rhesus monkeys (Macaca mulatta) detect rhythmic groups in music, but not the beat. *PLoS ONE, 7*, e51369.

Hoskins, J. (1993). *The play of time: Kodi perspectives on calendars, history, and exchange.* Berkeley: University of California Press.

Huang, Y., Jones, B. (1982). On the interdependence of temporal and spatial judgments. *Perception & Psychophysics, 32*, 7–14.

Hume, D. (1739/2000). *A treatise on human nature.* Oxford: Oxford University Press.

Hussain, F., Gupta, C., Hirning, A. J., Ott, W., Matthews, K. S., Josić K., Bennett, M. R. (2014). Engineered temperature compensation in a synthetic genetic clock. *Proceedings of the National Academy of Sciences, 111*, 972–977.

Huxley, T. H. (1894/1911). *Collected essays: Method and results.* New York: D. Appleton.

Ikeda, H., Kubo, T., Kuriyama, K., Takahashi, M. (2014). Self-awakening improves alertness in the morning and during the day after partial sleep deprivation. *Journal of Sleep Research, 23*, 673–680.

Ishihara, M., Keller, P. E., Rossetti, Y., Prinz, W. (2008). Horizontal spatial representations of time: Evidence for the STEARC effect. *Cortex, 44*, 454–461.

Ishizawa, Y., Ahmed, O. J., Patel, S. R., Gale, J. T., Sierra-Mercado, D., Brown, E. N., Eskandar, E. N. (2016). Dynamics of propofol-induced loss of consciousness across primate neocortex. *Journal of Neuroscience, 36*, 7718–7726.

Ivry, R. B., Schlerf, J. E. (2008). Dedicated and intrinsic models of time perception. *Trends in Cognitive Sciences, 12*, 273–280.

Jacobs, B., Schall, M., Prather, M., Kapler, E., Driscoll, L., Baca, S., Jacobs, J.,

Ford, K., Wainwright, M., Treml, M. (2001). Regional dendritic and spine variation in human cerebral cortex: A quantitative golgi study. *Cerebral Cortex, 11*, 558–571.

James, W. (1890). *The principles of psychology.* New York: Dover Publications.

Janssen, P., Shadlen, M. N. (2005). A representation of the hazard rate of elapsed time in the macaque area LIP. *Nature Neuroscience, 8*, 234–241.

Jazayeri, M., Shadlen, M. N. (2010). Temporal context calibrates interval timing. *Nature Neuroscience, 13*, 1020–1026.

————. (2015). A neural mechanism for sensing and reproducing a time interval. *Current Biology, 25*, 2599–2609.

Jin, D. Z., Fujii, N., Graybiel, A. M. (2009). Neural representation of time in cortico-basal ganglia circuits. *Proceedings of the National Academy of Science USA, 106*, 19156–19161.

Johnson, C. H., Golden, S. S., Kondo, T. (1998). Adaptive significance of circadian programs in cyanobacteria. *Trends in Microbiology, 6*, 407–410.

Johnson, H. A., Goel, A., Buonomano, D. V. (2010). Neural dynamics of in vitro cortical networks reflects experienced temporal patterns. *Nature Neuroscience, 13*, 917–919.

Jones, C. R., Campbell, S. S., Zone, S. E., Cooper, F., DeSano, A., Murphy, P. J., Jones, B., Czajkowski, L., Ptáček, L., J. (1999). Familial advanced sleep-phase syndrome: A short-period circadian rhythm variant in humans. *Nature Medicine, 5*, 1062–1065.

Jones, C. R., Huang, A. L., Ptáček, L. J., Fu, Y.-H. (2013). Genetic basis of human circadian rhythm disorders. *Experimental Neurology, 243*, 28–33.

Kable, J. W., Glimcher, P. W. (2007). The neural correlates of subjective value during intertemporal choice. *Nature Neuroscience, 10*, 1625–1633.

Kanabus, M., Szelag, E., Rojek, E., Poppel, E. (2002). Temporal order judgment for auditory and visual stimuli. *Acta Neurobiologiae Experimentalis, 62*, 263–270.

Kandel, E. (2013). The new science of mind and the future of knowledge. *Neuron, 80*, 546–560.

Kandel, E. R., Schartz, J., Jessel, T., Siegelbaum, S. A., Hudspeth, A. J. (2013). *Principles of neural science,* 5th ed. New York: McGraw-Hill Medical.

Karlsson, M. P., Frank, L. M. (2009). Awake replay of remote experiences in the hippocampus. *Nature Neuroscience, 12*, 913–918.

Karmarkar, U. R., Najarian, M. T., Buonomano, D. V. (2002). Mechanisms and significance of spike-timing dependent plasticity. *Biological Cybernetics, 87*, 373–382.

Keele, S. W., Pokorny, R. A., Corcos, D. M., Ivry, R. (1985). Do perception and motor production share common timing mechanisms: a correctional analysis. *Acta Psychologica (Amst.), 60*, 173–191.

Kiesel, A., Vierck, E. (2009). SNARC-like congruency based on number magnitude and response duration. *Journal of Experimental Psychology: Learning, Memory, and Cognition, 35,* 275–279.

Kilgard, M. P., Merzenich, M. M. (1995). Anticipated stimuli across skin. *Nature, 373,* 663.

———. (2002). Order-sensitive plasticity in adult primary auditory cortex. *Proceedings of the National Academy of Science USA, 99,* 3205–3209.

Killingsworth, M. A., Gilbert ,D. T. (2010). A wandering mind is an unhappy mind. *Science, 330,* 932.

Kim, J., Ghim, J.-W., Lee, J. H., Jung, M. W. (2013). Neural correlates of interval timing in rodent prefrontal cortex. *Journal of Neuroscience, 33,* 13834–13847.

Kivimäki, M., Batty, G. D., Hublin, C. (2011). Shift work as a risk factor for future type 2 diabetes: evidence, mechanisms, implications, and future research directions. *PLoS Med, 8,* e1001138.

Klampfl, S., David, S. V., Yin, P., Shamma, S. A., Maass, W. (2012). A quantitative analysis of information about past and present stimuli encoded by spikes of A1 neurons. *Journal of Neurophysiology, 108,* 1366–1380.

Knutsson, A. (2003). Health disorders of shift workers. *Occupational Medicine, 53,* 103–108.

Koch, C. (2004). *The quest for consciousness: A neurobiological approach.* Englewood, CO: Robers & Company.

Konopka, R. J., Benzer, S. (1971). Clock mutants of Drosophila melanogaster. *Proceedings of the National Academy of Science USA, 68,* 2112–2116.

Kording, K. (2007). Decision theory: What "should" the nervous system do? *Science, 318,* 606–610.

Kostarakos, K., Hedwig, B. (2012). Calling song recognition in female crickets: Temporal tuning of identified brain neurons matches behavior. *Journal of Neuroscience, 32,* 9601–9612.

Kraus, B. J., Robinson, R. J., White, J. A., Eichenbaum, H., Hasselmo, M. E. (2013). Hippocampal "time cells": Time versus path integration. *Neuron, 78,* 1090–1101.

Kwan, D., Craver, C. F., Green, L., Myerson, J., Boyer P., Rosenbaum, R. S. (2012). Future decision-making without episodic mental time travel. *Hippocampus, 22,* 1215–1219.

Kyriacou, C. P., Hall, J. C. (1980). Circadian rhythm mutations in Drosophila melanogaster affect short-term fluctuations in the male's courtship song. *Proceedings of the National Academy of Sciences of the USA, 77,* 6729–6733.

Laje, R., Buonomano, D. V. (2013). Robust timing and motor patterns by taming chaos in recurrent neural networks. *Nature Neuroscience 16,* 925–933.

Lakoff, G., Johnson, M. (1980/2003). *Metaphors we live by.* Chicago: University of Chicago Press.

Lamy, D., Salti, M., Bar-Haim ,Y. (2009). Neural correlates of subjective awareness and unconscious processing: an ERP study. *Journal of Cognitive Neuroscience*, *21*, 1435–1446.

Landes, D. S. (1983). *Revolution in time: Clocks and the making of the modern world.* New York: Barnes & Noble.

Lashley, K. S., ed. (1951). *The problem of serial order in behavior.* New York: Wiley.

Lasky, R. (2012). Time and the twin paradox. *Scientific American*, *21*, 30–33.

Lau, H. C., Rogers, R. D., Passingham R. E. (2007). Manipulating the experienced onset of intention after action execution. *Journal of Cognitive Neuroscience*, *19*, 81–90.

Lavie, P. (2001). Sleep-wake as a biological rhythm. *Annual Review of Psychology*, *52*, 277–303.

Lebedev, M. A., O'Doherty, J. E., Nicolelis, M. A. L. (2008). Decoding of temporal intervals from cortical ensemble activity. *Journal of Neurophysiology*, *99*, 166–186.

Lee, T. P., Buonomano, D. V. (2012). Unsupervised formation of vocalization-sensitive neurons: a cortical model based on short-term and homeostatic plasticity. *Neural Computation*, *24*, 2579–2603.

Lehiste, I. (1960). An acoustic–phonetic study of internal open juncture. *Phonetica*, *5 (suppl. 1)*, 5–54.

Lehiste, I., Olive, J. P., Streeter, L. A. (1976). Role of duration in disambiguating syntactically ambiguous sentences. *Journal of the Acoustical Society of America*, *60*, 1199–1202.

Leon, M. I., Shadlen, M. N. (2003). Representation of time by neurons in the posterior parietal cortex of the macaque. *Neuron*, *38*, 317–327.

Levine, R. (1996). *The geography of time.* New York: Basic Books.

Levy, W. B., Steward, O. (1983). Temporal contiguity requirements for long-term associative potentiation/depression in the hippocampus. *Neuroscience*, *8*, 791–797.

Lewis, P. A., Miall, R. C., Daan, S., Kacelnik, A. (2003). Interval timing in mice does not rely upon the circadian pacemaker. *Neuroscience Letters*, *348*, 131–134.

Libet, B., Gleason, C. A., Wright, E. W., Pearl, D. K. (1983). Time of conscious intention to act in relation to onset of cerebral activity (readiness-potential). The unconscious initiation of a freely voluntary act. *Brain*, *106 (Pt. 3)*, 623–642.

Lieving, L. M., Lane, S. D., Cherek, D. R., Tcheremissine, O. V. (2006). Effects of marijuana on temporal discriminations in humans. *Behavioral Pharmacology*, *17*, 173–183.

Livesey, A. C., Wall, M. B., Smith, A. T. (2007). Time perception: Manipulation of task difficulty dissociates clock functions from other cognitive demands. *Neuropsychologia*, *45*, 321–331.

Lockwood, M. (2005). *The labyrinth of time: Introducing the universe.* Oxford: Oxford University Press.

Loftus, E. F. (1996). *Eyewitness testimony*. Cambridge, MA: Harvard University Press.

Loftus, E. F., Schooler, J. W., Boone, S. M., Kline, D. (1987). Time went by so slowly: overestimation of event duration by males and females. *Applied Cognitive Psychology, 1*, 3–13.

Loh, D. H., Navarro, J., Hagopian, A., Wang, L. M., Deboer, T., Colwell, C. S. (2010). Rapid changes in the light/dark cycle disrupt memory of conditioned fear in mice. *PLoS ONE, 5*, e12546.

Lombardi, M. A. (2002). Fundamentals of time and frequency. In: *Mechanotronics handbook* (Bishop, RH, ed.). New York: CRC Press.

Long, M. A., Fee, M. S. (2008). Using temperature to analyse temporal dynamics in the songbird motor pathway. *Nature, 456*, 189–194.

Long, M. A., Jin, D. Z., Fee, M. S. (2010). Support for a synaptic chain model of neuronal sequence generation. *Nature, 468*, 394–399.

Maass, W., Natschläger, T., Markram, H. (2002). Real-time computing without stable states: A new framework for neural computation based on perturbations. *Neural Computation, 14*, 2531–2560.

MacDonald, C. J., Lepage, K. Q., Eden, U. T., Eichenbaum, H. (2011). Hippocampal "time cells" bridge the gap in memory for discontiguous events. *Neuron, 71*, 737–749.

MacDonald, C. J., Carrow, S., Place, R., Eichenbaum, H. (2013). Distinct hippocampal time cell sequences represent odor memories in immobilized rats. *Journal of Neuroscience, 33*, 14607–14616.

MacKillop, J., Amlung, M. T., Few, L. R., Ray, L. A., Sweet, L. H., Munafo, M. R. (2011). Delayed reward discounting and addictive behavior: a meta-analysis. *Psychopharmacology (Berl.), 216*, 305–321.

Mante, V., Sussillo, D., Shenoy, K. V., Newsome, W. T. (2013). Context-dependent computation by recurrent dynamics in prefrontal cortex. *Nature, 503*, 78–84.

Markram, H., Lubke, J., Frotscher, M., Sakmann, B. (1997). Regulation of synaptic efficacy by coincidence of postsynaptic APs and EPSPs. *Science, 275*, 213–215.

Martin, F. H., Garfield, J. (2006). Combined effects of alcohol and caffeine on the late components of the event-related potential and on reaction time. *Biological Psychology, 71*, 63–73.

Matell, M. S., Meck, W. H. (2004). Cortico-striatal circuits and interval timing: coincidence detection of oscillatory processes. *Cognitive Brain Research 21*, 139–170.

Matsuda, F. (1996). Duration, distance, and speed judgments of two moving objects by 4- to 11-year olds. *Journal of Experimental Child Psychology, 63*, 286–311.

Matthews, M. R. (2000). *Time for science education: how teaching the history and philosophy of pendulum motion can contribute to science literacy.* New York: Kluwer Academic.

Matthews, W. J. (2015). Time perception: The surprising effects of surprising stimuli. *Journal of Experimental Psychology: General, 144,* 172–197.

Matthews, W. J., Meck, W. H. (2016). Temporal cognition: Connecting subjective time to perception, attention, and memory. *Psychological Bulletin, 142,* 865–890.

Mauk, M. D., Donegan, N. H. (1997). A model of Pavlovian eyelid conditioning based on the synaptic organization of the cerebellum. *Learning & Memory, 3,* 130–158.

Mauk, M. D., Buonomano, D. V. (2004). The neural basis of temporal processing. *Annual Review of Neuroscience, 27,* 307–340.

McClure, G. Y., McMillan, D. E. (1997). Effects of drugs on response duration differentiation. VI: differential effects under differential reinforcement of low rates of responding schedules. *Journal of Pharmacology and Experimental Therapeutics, 281,* 1368–1380.

McClure, S. M., Laibson, D. I., Loewenstein, G., Cohen, J. D. (2004). Separate neural systems value immediate and delayed monetary rewards. *Science, 306,* 503–507.

McGlone, M. S., Harding, J. L. (1998). Back (or forward?) to the future: The role of perspective in temporal language comprehension. *Journal of Experimental Psychology: Learning, Memory, and Cognition, 24,* 1211–1223.

Meck, W. H. (1996). Neuropharmacology of timing and time perception. *Cognitive Brain Research, 3,* 227–242.

Medina, J. F., Garcia, K. S., Nores, W. L., Taylor, N. M., Mauk, M. D. (2000). Timing mechanisms in the cerebellum: testing predictions of a large-scale computer simulation. *Journal of Neuroscience, 20,* 5516–5525.

Mégevand, P., Molholm, S., Nayak, A., Foxe, J. J. (2013). Recalibration of the multisensory temporal window of integration results from changing task demands. *PLoS ONE, 8,* e71608.

Meijer, J. H., Robbers ,Y. (2014). Wheel running in the wild. *Proceedings of the Royal Society of London B: Biological Sciences, 281,* 20140210.

Mello, G. B. M., Soares, S., Paton, J. J. (2015). A scalable population code for time in the striatum. *Current Biology, 9,* 1113–1122.

Merchant, H., Harrington, D. L., Meck, W. H. (2013). Neural basis of the perception and estimation of time. *Annual Review of Neuroscience, 36,* 313–336.

Meyer, L. (1961). *Emotion and meaning in music.* Chicago: University of Chicago Press.

Miall, C. (1989). The storage of time intervals using oscillating neurons. *Neural Computation, 1,* 359–371.

Milham, W. I. (1941) *Time & timekeepers: Including the history, construction, care, and accuracy of clocks and watches.* London: Macmillan, 37

Mita, A., Mushiake, H., Shima, K., Matsuzaka, Y., Tanji, J. (2009). Interval time coding by neurons in the presupplementary and supplementary motor areas. *Nature Neuroscience, 12,* 502–507.

Modi, M. N., Dhawale, A. K., Bhalla, U. S. (2014). CA1 cell activity sequences emerge after reorganization of network correlation structure during associative learning. *Elife, 3,* e01982.

Montague, P. R. (2008). Free will. *Current Biology, 18,* R584–R585.

Moorcroft, W. H., Kayser, K. H., Griggs, A. J. (1997). Subjective and objective confirmation of the ability to self-awaken at a self-predetermined time without using external means. *Sleep, 20,* 40–45.

Morrone, M. C., Ross, J., Burr, D. (2005). Saccadic eye movements cause compression of time as well as space. *Nature Neuroscience, 8,* 950–954.

Morrow, N. S., Schall, M., Grijalva, C. V., Geiselman, P. J., Garrick, T., Nuccion, S., Novin, D. (1997). Body temperature and wheel running predict survival times in rats exposed to activity-stress. *Physiology & Behavior, 62,* 815–825.

Muller, T., Nobre, A. C. (2014). Perceiving the passage of time: neural possibilities. *Annals of the New York Academy of Sciences, 1326,* 60–71.

Mumford, L. (2010/1934). *Technics & civilization.* Chicago: University of Chicago Press.

Murakami, M., Vicente, M. I., Costa, G. M., Mainen, Z. F. (2014). Neural antecedents of self-initiated actions in secondary motor cortex. *Nature Neuroscience, 17,* 1574–1582.

Nagarajan, S. S., Blake, D. T., Wright, B. A., Byl, N., Merzenich, M. M. (1998). Practice-related improvements in somatosensory interval discrimination are temporally specific but generalize across skin location, hemisphelre, and modality. *Journal of Neuroscience, 18,* 1559–1570.

Nichols, S. (2011). Experimental philosophy and the problem of free will. *Science, 331,* 1401–1403.

Nikolić, D., Häusler, S., Singer, W., Maass, W. (2009). Distributed fading memory for stimulus properties in the primary visual cortex. *PLoS Biol, 7,* e1000260.

Noyes, R., Kletti, R. (1972). The experience of dying from falls. *Omega, 3,* 45–52.

———. (1976). Depoersonalization in face of life-threatening danger. *Psychiatry-Interpersonal and Biological Processes, 39,* 19–27.

Núñez, R., Cooperrider, K. (2013). The tangle of space and time in human cognition. *Trends in Cognitive Science, 17,* 220–229.

Núñez, R. E., Sweetser, E. (2006). With the future behind them: convergent evidence from Aymara language and gesture in the crosslinguistic comparison of spatial construals of time. *Cognitive Science, 30,* 401–450.

Ohyama, T., Nores, W. L., Murphy, M., Mauk, M. D. (2003). What the cerebellum computes. *Trends in Neurosciences, 26*, 222–227.

Oliveri, M., Vicario, C. M., Salerno, S., Koch, G., Turriziani, P., Mangano, R., Chillemi, G., Caltagirone, C. (2008). Perceiving numbers alters time perception. *Neuroscience Letters, 438*, 308–311.

Ornstein, R. E. (1969). *On the experience of time.* Harmondsworth: Penguin.

Osvath, M., Persson, T. (2013). Great apes can defer exchange: a replication with different results suggesting future oriented behavior. *Frontiers in Psychology, 4*, 698.

Ouyang, Y., Andersson, C. R., Kondo, T., Golden, S. S., Johnson, C. H. (1998). Resonating circadian clocks enhance fitness in cyanobacteria. *Proceedings of the National Academy of Sciences, 95*, 8660–8664.

Panagiotaropoulos, T. I., Deco, G., Kapoor, V., Logothetis, N. K. (2012). Neuronal discharges and gamma oscillations explicitly reflect visual consciousness in the lateral prefrontal cortex. *Neuron, 74*, 924–935.

Papachristos, E. B., Jacobs, E. H., Elgersma, Y. (2011). Interval timing is intact in arrhythmic Cry1/Cry2-deficient mice. *Journal of Biological Rhythms, 26*, 305–313.

Papert, S. (1999). Child psychologist Jean Piaget. *Time*, March 29.

Pariyadath, V., Eagleman, D. M. (2007). The effect of predictability on subjective duration. *PLoS ONE, 2*, e1264.

Park, J., Schlag-Rey, M., Schlag, J. (2003). Voluntary action expands perceived duration of its sensory consequence. *Experimental Brain Research, 149*, 527–529.

Pastalkova, E., Itskov, V., Amarasingham, A., Buzsaki, G. (2008). Internally generated cell assembly sequences in the rat hippocampus. *Science, 321*, 1322–1327.

Patel, A. B., Loerwald, K. W., Huber, K. M., Gibson, J. R. (2014). Postsynaptic FMRP promotes the pruning of cell-to-cell connections among pyramidal neurons in the L5A neocortical network. *Journal of Neuroscience, 34*, 3413–3418.

Patel, A. D. (2006). Musical rhythm, linguistic rhythm, and human evolution. *Music Perception, 24*, 99–104.

Patel, A. D., Iversen, J. R., Bregman, M. R., Schulz, I. (2009). Studying synchronization to a musical beat in nonhuman animals. *Annals of the New York Academy of Sciences, 1169*, 459–469.

Penrose, R. (1989/1999). *The emperor's new mind.* Oxford: Oxford University Press.

Perrett, S. P., Ruiz, B. P., Mauk, M. D. (1993). Cerebellar cortex lesions disrupt learning-dependent timing of conditioned eyelid responses. *Journal of Neuroscience, 13*, 1708–1718.

Peters, J. (2011). The role of the medial orbitofrontal cortex in intertemporal choice: Prospection or valuation? *Journal of Neuroscience, 31,* 5889–5890.

Peters, J., Büchel, C. (2010). Episodic future thinking reduces reward delay discounting through an enhancement of prefrontal-mediotemporal interactions. *Neuron, 66,* 138–148.

Piaget, J. (1946/1969). *The child's conception of time.* New York: Basic Books.

———. (1972). *Psychology and epistemology: towards a theory of knowledge.* London: Penguin Press.

Pierce, W. D., Epling, W. F., Boer, D. P. (1986). Deprivation and satiation: The interrelations between food and wheel running. *Journal of the Experimental Analysis of Behavior, 46,* 199–210.

Pinker, S. (2007). *The stuff of thought: Language as a window into human nature.* New York: Penguin Books.

———. (2014). *The sense of style: The thinking person's guide to writing in the 21st century.* New York: Penguin.

Popper, K. (1992). *Unended quest: An intellectual autobiography.* London: Routledge.

Prelec, D., Simester, D. (2001). Always leave home without it: A further investigation of the credit-card effect on willingness to pay. *Marketing Letters, 12,* 5–12.

Price-Williams, D. R. (1954). The kappa effect. *Nature, 173,* 363–364.

Prigogine, I., Stengers, I. (1984). *Order out of chaos: man's new dialogue with nature.* Toronto: Bantam.

Purdon, P. L., Pierce, E. T., Mukamel, E. A., Prerau, M. J., Walsh, J. L., Wong, K. F. K., Salazar-Gomez, A. F., Harrell, P. G., Sampson A. L., Cimenser, A., Ching, S., Kopell, N. J., Tavares-Stoeckel, C., Habeeb, K., Merhar, R., Brown, E. N. (2013). Electroencephalogram signatures of loss and recovery of consciousness from propofol. *Proceedings of the National Academy of Sciences, 110,* E1142–E1151.

Purves, D., Brannon, E. M., Cabeza, R., Huettel, S. A., LaBar, K. S., Platt, M. L., Woldorff, M. G. (2008). *Principles of Cognitive Neuroscience.* Sunderland, MA: Sinauer.

Putnam, H. (1967). Time and physical geometry. *Journal of Philosophy, 64,* 240–247.

Quintana, J., Fuster, J. M. (1992). Mnemonic and predictive functions of cortical neurons in a memory task. *Neuroreport, 3,* 721–724.

Raby, C. R., Alexis, D. M., Dickinson, A., Clayton, N. S. (2007). Planning for the future by western scrub-jays. *Nature, 445,* 919–921.

Race, E., Keane, M. M., Verfaellie, M. (2011). Medial temporal lobe damage causes deficits in episodic memory and episodic future thinking not attributable to deficits in narrative construction. *Journal of Neuroscience, 31,* 10262–10269.

Rae, A. (1986). *Quantum physics: illusion or reality?* Cambridge: Cambridge University Press.

Raghubir, P., Srivastava, J. (2008). Monopoly money: the effect of payment coupling and form on spending behavior. *Journal of Experimental Psychology Applied, 14*, 213–225.

Ralph, M. R., Foster, R. G., Davis, F. C., Menaker, M. (1990). Transplanted suprachiasmatic nucleus determines circadian period. *Science, 247*, 975–978.

Rammsayer, T. (1992). Effects of benzodiazepine-induced sedation on temporal processing. *Human Psychopharmacology, 7*, 311–318.

Rammsayer, T. H. (1999). Neuropharmacological evidence for different timing mechanisms in humans. *Quarterly Journal of Experimental Psychology B, 52*, 273–286.

Rammsayer, T. H., Vogel, W. H. (1992). Pharmacological properties of the internal clock underlying time perception in humans. *Neuropsychobiology, 26*, 71–80.

Rammsayer, T. H., Buttkus, F., Altenmuller, E. (2012). Musicians do better than nonmusicians in both auditory and visual timing tasks. *Music Perception, 30*, 85–96.

Raymond, J., Lisberger, S. G., Mauk, M. D. (1996). The cerebellum: a neuronal learning machine? *Science, 272*, 1126–1132.

Reddy, P., Zehring, W. A., Wheeler, D. A., Pirrotta, V., Hadfield, C., Hall, J. C., Rosbash, M. (1984). Molecular analysis of the period locus in Drosophila melanogaster and identification of a transcript involved in biological rhythms. *Cell, 38*, 701–710.

Rennaker, R. L., Carey, H. L., Anderson, S. E., Sloan, A. M., Kilgard, M. P. (2007). Anesthesia suppresses nonsynchronous responses to repetitive broadband stimuli. *Neuroscience, 145*, 357–369.

Reyes, A., Sakmann, B. (1999). Developmental switch in the short-term modification of unitary EPSPs evoked in layer 2/3/ and layer 5 pyramidal neurons of rat neocortex. *Journal of Neuroscience 19*, 3827–3835.

Richards, W. (1973). Time reproductions by H.M. *Acta Psychologica, 37*, 279–282.

Rietdijk, C. W. (1966). A rigorous proof of determinism derived from the special theory of relativity. *Philosophy of Science, 33*, 341–344.

Rigotti, M., Barak, O., Warden, M. R., Wang, X.-J., Daw, N. D., Miller, E. K., Fusi, S. (2013). The importance of mixed selectivity in complex cognitive tasks. *Nature, 497*, 585–590.

Rose, G., Leary, C., Edwards, C. (2011). Interval-counting neurons in the anuran auditory midbrain: factors underlying diversity of interval tuning. *Journal of Comparative Physiology A: Neuroethology, Sensory, Neural, and Behavioral Physiology, 197*, 97–108.

Routtenberg, A., Kuznesof, A. W. (1967). Self-starvation of rats living in activity wheels on a restricted feeding schedule. *Journal of Comparative and Physiological Psychology, 64*, 414–421.

Rovelli, C. (2004). *Quantum gravity.* Cambridge: Cambridge University Press.

Sacks, O. (2004). Speed. *The New Yorker*, Aug. 23, 60–69.

Sadagopan, S., Wang, X. (2009). Nonlinear spectrotemporal interactions underlying selectivity for complex sounds in auditory cortex. *Journal of Neuroscience, 29*, 11192–11202.

Saj, A., Fuhrman, O., Vuilleumier, P., Boroditsky, L. (2014). Patients with left spatial neglect also neglect the "left side" of time. *Psychological Science, 25*, 207–214.

Salti, M., Monto, S., Charles, L., King, J.-R., Parkkonen, L., Dehaene, S. (2015). Distinct cortical codes and temporal dynamics for conscious and unconscious percepts. *eLife, 4*, e05652.

Sarrazin, J. C., Giraudo, M. D., Pailhous, J., Bootsma, R. J. (2004). Dynamics of balancing space and time in memory: tau and kappa effects revisited. *Journal of Experimental Psychology Human Perception and Performance, 30*, 411–430.

Sauer, T., ed. (2014). *Piaget, Einstein, and the concept of time.* Berlin: Edition Open Access.

Scarf, D., Smith, C., Stuart, M. (2014). A spoon full of studies helps the comparison go down: a comparative analysis of Tulving's spoon test. *Frontiers in Psychology, 5*, 893.

Schacter, D. L. (1996). *Searching for memory.* New York: Basic Books.

Schacter, D. L., Addis, D. R. (2007). Constructive memory: The ghosts of past and future. *Nature, 445*, 27.

Schacter, D. L., Addis D. R., Buckner R. L. (2007). Remembering the past to imagine the future: the prospective brain. *Nature Reviews Neuroscience, 8*, 657–661.

Scharnowski, F., Rüter, J., Jolij, J., Hermens, F., Kammer T., Herzog, M. H. (2009). Long-lasting modulation of feature integration by transcranial magnetic stimulation. *Journal of Vision, 9*, 1–10.

Schuster, S., Rossel, S., Schmidtmann, A., Jäger, I., Poralla, J. (2004). Archer fish learn to compensate for complex optical distortions to determine the absolute size of their aerial prey. *Current Biology, 14*, 1565–1568.

Schwab, S., Miller, J. L., Grosjean, F., Mondini, M. (2008). Effect of speaking rate on the identification of word boundaries. *Phonetica, 65*, 173–186.

Seeyave, D. M., Coleman, S., Appugliese, D., Corwyn, R. F., Bradley, R. H., Davidson, N. S., Kaciroti, N., Lumeng, J. C. (2009). ability to delay gratification at age 4 years and risk of overweight at age 11 years. *Archives of Pediatric & Adolescent Medicine, 163*, 303–308.

Sellitto, M., Ciaramelli, E., di Pellegrino, G. (2010). Myopic discounting of future rewards after medial orbitofrontal damage in humans. *Journal of Neuroscience, 30*, 16429–16436.

Sergent, C., Wyart, V., Babo-Rebelo, M., Cohen, L., Naccache, L., Tallon-Baudry, C. (2013). Cueing attention after the stimulus is gone can retrospectively trigger conscious perception. *Current Biology, 23*, 150–155.

Sewell, R. A., Schnakenberg, A., Elander, J., Radhakrishnan, R., Williams, A., Skosnik, P. D., Pittman,, B., Ranganathan, M., D'Souza, D. C. (2013). Acute effects of THC on time perception in frequent and infrequent cannabis users. *Psychopharmacology, 226*, 401–413.

Shariff, A. F., Vohs, K. D. (2014). The world without free will. *Scientific American*, June, 77–79.

Sharma, V. K. (2003). Adaptive significance of circadian clocks. *Chronobiology International, 20*, 901–919.

Shepherd, G. M. (1998). *The synaptic organization of the brain*. New York: Oxford University.

Shuler, M. G., Bear, M. F. (2006). Reward timing in the primary visual cortex. *Science, 311*, 1606–1609.

Siegler, R. S., Richards, D. D. (1979). Development of time, speed, and distance concepts. *Developmental Psychology, 15*, 288–298.

Smart, J. J. C., ed. (1964). *Problems of space and time*. New York: Macmillan.

Smith, K. (2011). Neuroscience vs philosophy: taking aim at free will. *Nature, 477*, 23–25.

Smolen, P., Hardin, P. E., Lo, B. S., Baxter, D. A., Byrne, J. H. (2004). Simulation of Drosophila circadian oscillations, mutations, and light responses by a model with VRI, PDP-1, and CLK. *Biophysical Journal, 86*, 2786–2802.

Smolin, L. (2013). *Time reborn: from the crises in physics to the future of the universe*. New York: Houghton Mifflin Harcourt.

Sompolinsky, H., Crisanti, A., Sommers, H. J. (1988). Chaos in random neural networks. *Physical Review Letters, 61*, 259–262.

Soon, C. S., Brass, M., Heinze, H.-J., Haynes, J.-D. (2008). Unconscious determinants of free decisions in the human brain. *Nature Neuroscience, 11*, 543–545.

Spalding, Kirsty L., Bergmann, O., Alkass, K., Bernard, S., Salehpour, M., Huttner, H. B., Boström, E., Westerlund, I., Vial, C., Buchholz, B. A., Possnert, G., Mash, D. C., Druid, H., Frisén, J. (2013). Dynamics of hippocampal neurogenesis in adult humans. *Cell, 153*, 1219–1227.

Stern, D. L. (2014). Reported Drosophila courtship song rhythms are artifacts of data analysis. *BMC Biology, 12*, 38.

Stetson, C., Fiesta, M. P., Eagleman, D. M. (2007). Does time really slow down during a frightening event? *PLoS ONE, 2*, e1295.

Stokes, Mark G., Kusunoki, M., Sigala N., Nili, H., Gaffan, D., Duncan, J.

(2013). Dynamic coding for cognitive control in prefrontal cortex. *Neuron*, *78*, 364–375.

Suddendorf, T. (2013). *The gap: the science of what separates us from other animals.* New York: Basic Books.

Suddendorf, T., Corballis, M. C. (1997). Mental time travel and the evolution of the human mind. *Genetic, Social, and General Psychology Monographs*, *123*, 133–167.

———. (2007). The evolution of foresight: What is mental time travel, and is it unique to humans? *Behavior and Brain Sciences*, *30*, 299–313; discussion 313–351.

Summa, K. C., Turek, F. W. (2015). The clocks within us. *Scientific American*, January, 50–55.

Surowiecki, J. (2013). Deadbeat governments. *The New Yorker*, Dec. 23, 46.

Sussillo, D., Barak, O. (2013). Opening the black box: Low-dimensional dynamics in high-dimensional recurrent neural networks. *Neural Computation*, *25*, 626–649.

Swann, A. C., Lijffijt, M., Lane, S. D., Cox, B., Steinberg, J. L., Moeller, F. G. (2013). Norepinephrine and impulsivity: effects of acute yohimbine. *Psychopharmacology*, *229*, 83–94.

Taler, V., Baum, S. R., Chertkow, H., Saumier, D. (2008). Comprehension of grammatical and emotional prosody is impaired in Alzheimer's disease. *Neuropsychology*, *22*, 188–195.

Taylor, M., Esbensen, B. M., Bennett, R. T. (1994). Children's understanding of knowledge acquisition: the tendency for children to report that they have always known what they have just learned. *Child Development*, *65*, 1581–1604.

Terry, P., Doumas, M., Desai, R. I., Wing, A. M. (2008). Dissociations between motor timing, motor coordination, and time perception after the administration of alcohol or caffeine. *Psychopharmacology (Berl.).*

Tessmar-Raible, K., Raible, F., Arboleda, E. (2011). Another place, another timer: Marine species and the rhythms of life. *BioEssays*, *33*, 165–172.

Thorne, K. S. (1995). *Black holes and time warps: Einstein's outrageus legacy.* New York: W. W. Norton.

Tinklenberg, J. R., Roth, W. T., Kopell, B. S. (1976). Marijuana and ethanol: Differential effects on time perception, heart rate, and subjective response. *Psychopharmacology*, *49*, 275–279.

Toh, K. L., Jones, C. R., He, Y., Eide, E. J., Hinz, W. A., Virshup, D. M., Ptáček, L. J., Fu, Y.-H. (2001). An hPer2 phosphorylation site mutation in familial advanced sleep phase syndrome. *Science*, *291*, 1040–1043.

Toida, K., Ueno, K., Shimada, S. (2014). Recalibration of subjective simultaneity between self-generated movement and delayed auditory feedback. *Neuroreport*, *25*, 284–288.

Tom, G., Burns, M., Zeng, Y. (1997). Your life on hold: The effect of telephone waiting time on customer perception. *Journal of Interactive Marketing, 11,* 25–31.

Treisman, M. (1963). Temporal discrimination and the indifference interval: implications for a model of the 'internal clock.' *Psychological Monographs, 77,* 1–31.

Tse, P. U., Intriligator, J., Rivest, J., Cavanagh, P. (2004). Attention and the subjective expansion of time. *Perception & Psychophysics, 66,* 1171–1189.

Tulving, E. (1985). Memory and consciousness. *Canadian Psychologist, 26,* 1–12.

———, ed. (2005). *Episodic memory and autonoesis: Uniquely human?* New York: Oxford University Press.

Van der Burg, E., Alais, D., Cass, J. (2015). Audiovisual temporal recalibration occurs independently at two different time scales. *Scientific Reports, 5,* 14526.

Van Wassenhove, V. (2009). Minding time in an amodal representational space. *Philosophical Transactions of the Royal Society of London B Biological Science, 364,* 1815–1830.

Van Wassenhove, V., Grant, K. W., Poeppel, D. (2007). Temporal window of integration in auditory-visual speech perception. *Neuropsychologia, 45,* 598–607.

Vitaterna, M. H., King, D. P., Chang, A.-M., Kornhauser, J. M., Lowrey, P. L., McDonald, J. D., Dove, W. F., Pinto, L. H., Turek, F. W., Takahashi, J. S. (1994). Mutagenesis and mapping of a mouse gene, clock, essential for circadian behavior. *Science* (New York), *264,* 719–725.

Walsh, V. (2003). A theory of magnitude: common cortical metrics of time, space and quantity. *Trends in Cognitive Sciences, 7,* 483–488.

Wearden, J. H. (2015). Passage of time judgments. *Consciousness and Cognition, 38,* 165–171.

Wearden, J. H., Edwards, H., Fakhri, M., Percival, A. (1998). Why "sounds are judged longer than lights": Application of a model of the internal clock in humans. *Quarterly Journal of Experimental Psychology Section B, 51,* 97–120.

Wearden, J. H., O'Donoghue, A., Ogden, R., Montgomery, C. (2014). Subjective duration in the laboratory and the world outside. In: *Subjective time: The philosophy, psychology, and neuroscience of temporality* (Arstila, V, Lloyd, D, eds.), 287–306. Cambridge, MA: MIT Press.

Weaver, D. R. (1998). The suprachiasmatic nucleus: A 25-year retrospective. *Journal of Biological Rhythms, 13,* 100–112.

Wegner, D. M. (2002). *The illusion of conscious will.* Cambridge, MA: MIT Press.

Weiner, J. (1999). *Time, love, memory: A great biologist and his quest for the origins of behavior.* New York: Vintage Books.

Wells, R. B. D. (1860). *Illustrated hand-book of phrenology, physiology, and physiognomy.* London: H. Vickers.

Welsh, D., Engle, E. R. A., Richardson, G., Dement, W. (1986). Precision of circadian wake and activity onset timing in the mouse. *Journal of Comparative Physiology A, 158,* 827–834.

Weyl, H. (1949/2009). *Philosophy of mathematics and natural science.* Princeton: Princeton University Press.

Whiting, A., Donthu N. (2009). Closing the gap between perceived and actual waiting times in a call center: Results from a field study. *Journal of Services Marketing, 23,* 279–288.

Wiener, M., Turkeltaub, P., Coslett, H. B. (2010). The image of time: A voxel-wise meta-analysis. *Neuroimage, 49,* 1728–1740.

Wilson, M. A., McNaughton, B. L. (1994). Reactivation of hippocampal ensemble memories during sleep. *Science, 265,* 676–679.

Wise, S. P. (2008). Forward frontal fields: phylogeny and fundamental function. *Trends in Neurosciences, 31,* 599–608.

Wiskott, L., Sejnowski, T. J. (2002). Slow feature analysis: unsupervised learning of invariances. *Neural Computation, 14,* 715–770.

Wittmann, M., Paulus, M. P. (2007). Decision making, impulsivity and time perception. *Trends Cognitive Sciences, 12,* 7–12.

Wood, J. N., Grafman, J. (2003). Human prefrontal cortex: processing and representational perspectives. *Nature Reviews Neuroscience, 4,* 139–147.

Wright, B. A., Buonomano, D. V., Mahncke, H. W., Merzenich, M. M. (1997). Learning and generalization of auditory temporal-interval discrimination in humans. *Journal of Neuroscience, 17,* 3956–3963.

Wright, B. A., Wilson, R. M., Sabin, A. T. (2010). Generalization lags behind learning on an auditory perceptual task. *Journal of Neuroscience, 30,* 11635–11639.

Xuan, B., Zhang, D., He, S., Chen, X. (2007). Larger stimuli are judged to last longer. *Journal of Vision, 7,* 1–5.

Yang, Y., Duguay, D., Bédard, N., Rachalski, A., Baquiran, G., Na, C. H., Fahrenkrug J., Storch, K.-F., Peng, J., Wing, S. S., Cermakian, N. (2012). Regulation of behavioral circadian rhythms and clock protein PER1 by the deubiquitinating enzyme USP2. *Biology Open 1:*789–801.

Yarrow, K., Haggard, P., Heal, R., Brown, P., Rothwell, J. C. (2001). Illusory perceptions of space and time preserve cross-saccadic perceptual continuity. *Nature, 414,* 302–305.

Zakay, D., Block, R. A. (1997). Temporal cognition. *Current Directions in Psychological Science, 6,* 12–16.

Zantke, J., Ishikawa-Fujiwara, T., Arboleda, E., Lohs, C., Schipany, K., Hallay, N., Straw, A. D., Todo, T., Tessmar-Raible, K. (2013). Circadian and circalunar clock interactions in a marine annelid. *Cell Reports.*

Zarco, W., Merchant, H., Prado, L., Mendez, J. C. (2009). Subsecond timing in primates: comparison of interval production between human subjects and rhesus monkeys. *Journal of Neurophysiology, 102*, 3191–3202.

Zeh, H. D. (1989/2007). *The physical basis of the direction of time.* Berlin: Springer.

Zhou, X., de Villers-Sidani, É., Panizzutti, R., Merzenich, M. M. (2010). Successive-signal biasing for a learned sound sequence. *Proceedings of the National Academy of Sciences, 107*, 14839–14844.

Zimbardo, P., Boyd, J. (2008). *The time paradox.* New York: Free Press.

Zucker, R. S. (1989). Short-term synaptic plasticity. *Annual Review of Neuroscience, 12*, 13–31.

Zucker, R. S., Regehr, W. G. (2002). Short-term synaptic plasticity. *Annual Review of Physiology, 64*, 355–405.

INDEX

Page numbers followed by *n* refer to footnotes.
Page numbers in *italics* refer to figures and illustrations.

accuracy, 36
A Christmas Carol (Dickens), 18
action potentials (spikes), *28*, 69, 94,
    99, 226
active state (neural networks), 110,
    112, 227
actograms, 35, *36*
Adams, Douglas, 215
adult neurogenesis, 129–30, 245*n*
El Anacronópete (Gaspard), 238*n*
animals
    hardwired instincts, 199–200
    mental time travel, 199–202, 213
    navigation through space, 5, 182
    senses and space, 5–6, 182, 237*n*
Anne of England (queen), 137
anterograde amnesia, 197–98
apes, mental time travel, 201–3, 213
arrow of time
    entropy and, 151–55
    time-irreversible processes
        ("arrows") in quantum mechan-
        ics, 155–56
    time-symmetric equations of phys-
        ics, 150–51, 246*n*
    see also flow of time
atomic bomb testing, 129–30

atomic clocks, 52, 127–28, 140–41,
    143, 161, 246*n*
auditory cortex, 97, 210, 253*n*
Augustine (saint), 4, 16, 222

Babylonian calendar, 133
Back to the Future (movie), 17
backward editing in time, 219–20,
    222
Barbour, Julian, 8, 171, 254*n*
Baron-Cohen, Sacha, 81–82
basal ganglia, 96
basketball game thought experiment,
    157–58, 247*n*
beat
    as alignment of biological oscilla-
        tors, 104
    animal inability to keep a beat,
        90–91
    defined, 90, 104
    prediction in keeping a beat, 90, 92
    vocal learning hypothesis, 91–92
    see also interval discrimination
Benzer, Seymour, 45, 50
Bible, 17
Big Bang
    low-entropy state after, 155

bimetallic strips, 138
bipolar disorder, 225
birds
  ability to keep a beat, 90–92
  HVC neurons, 93–95, 111
  scrub jays, 200–201, 213
  songbirds, 92–95, 111, 119
  synfire chain, *94*
  vocal learning, 91–92
  zebra finches, 92–93
blackjack, 26–27, 239*n*
block universe
  acceptance in physics and philoso-
    phy, 156
  block-universe/time-flow paradox,
    170–72, 175
  challenges and failings of, 230–31
  compatibility with neuroscience,
    172–76
  consciousness and neural dynamics
    in, 175–76, 249*n*
  Einstein acceptance of, 249*n*
  eternalism, 168–69
  under eternalism, 12, 13–14, 27,
    146–47
  evolving block universe theory, 147,
    228
  free will and, 224
  moments-within-a-moment
    hypothesis, 172, 175, 176, 249*n*
  special relativity, 19, 149, 248–
    49*n*
  see also spacetime; spatialization of
    time
body awareness, 72–73, 78, 174
Boltzmann, Ludwig, 151–52, 155
bonobos, 201
*Borat: Cultural Learnings of America
    for Make Benefit Glorious Nation
    of Kazakhstan* (movie), 82
Borges, Jorge Luis, 195
Boroditsky, Lera, 187, 189
brain
  auditory cortex, 97, 210, 253*n*
  brain dynamics, 116–18

cerebellum and motor timing, 96,
    112, 115
  consciousness and neural dynamics,
    73–74, 175–76, 249*n*
  encephalization quotient, 210
  inherent limitations and biases,
    145–46
  internal dynamics of neural net-
    work, 97–98
  mental time travel in, 209–13
  nervous system attuned to laws of
    physics, 173, 178
  number of neurons in brain, 27,
    109
  parietal cortex and information
    about quantity, 188, 189
  prefrontal cortex, 210–11, 213,
    252*n*
  primate brain evolution, 144–45
  ratio of brain to body mass, 209–10
  supplementary motor area, 96, 225
  see also suprachiasmatic nucleus
brain as time machine
  mental travel back and forth in
    time, 22–23
  perceiving the passage of time,
    21–22
  remembering past to predict future,
    10–11, 20, 90, 232
  telling time, 21, 32, 232
Brain Bugs (Buonomano), 26
brain stem, 67
Bundy, Willard, 141–42
Buonomano, Dean
  blackjack experiences, 26–27
  experience of the slow-motion
    effect, 72
butterfly effect, 8, 119

Caesar, Julius, 133
caffeine, 70
calendars, 132–33
Callender, Craig, 150
cannabinoids, 67, 68
carbon radioisotope (^{14}C), 129–31

Carnap, Rudolf, 169–70
cause and effect
  rules of, 23–25, 27
  synaptic cause and effect, 27–31
  temporal asymmetry of, 23, 25, 27, 29
  temporal misdirection, 26–27
cell division and circadian clocks, 45, 239n
cellular oscillations, 42
cerebellum and motor timing, 96, 112, 115
cesium 133 resonance frequency, 140–41, 148
chaos
  butterfly effect in weather, 8, 119
  computer models of neural networks, 118–24
  defined, 119
  logistic equation, 119–20
  word written by recurrent network simulation, 122, 123
  see also recurrent neural networks
child-on-a slide timer, 105
children and time, 179, 180–81
The Child's Conception of Time (Piaget), 179
chimpanzees, 201–2, 210
chronobiology, 10
chronopharmacology, 64–68
chronostasis, 58, 63
chronotypes, 50
circadian clocks
  accuracy, 36–37, 136, 246n
  cell division and, 45, 239n
  cellular oscillations, 42
  circadian rhythm sleep disorders, 50
  cyanobacteria, 43–44
  in diurnal animals, 37
  escape from light hypothesis, 45, 239n
  fighting the clock, 50–52
  as free-running rhythms, 37–38, 40, 41, 46, 52, 55

hidden from conscious access, 55
  in individual cells, 41–42, 43–44
  isolation experiments, 37–40, 50, 52
  jet lag, 48–49, 53
  mechanics of, 45–47, 103
  in mice, 35–37
  Mimosa pudica ("touch-me-not"), 41
  in nocturnal animals, 37
  photosynthesis and, 43–44
  self-awakening by humans, 37
  shift workers, 51
  sleep-wake cycles, 35–40, 50, 53, 136
  temperature compensation, 47, 240n
  temporal judgments and, 42
  vs. timing on the scale of seconds, 53, 240n
  transcription/translation autoregulatory feedback loop, 46–47, 53, 102
  see also suprachiasmatic nucleus; telling time
circadian rhythm sleep disorders, 50
circalunar cycles, 53–55
circular time, 238n
classical conditioning, 24–25, 115, 238n
Clayton, Nicola, 200–201
clepsydra (water clocks), 134–35
climate change, 206
clock punching, 142
clocks
  atomic clocks, 52, 127–28, 140–41, 143, 161, 246n
  clepsydra (water clocks), 134–35
  defined, 15
  electrocoordination, 139
  industrial revolution and, 43
  marine chronometers, 138–39
  mechanical clocks, invention, 135
  oscillators in, 48, 101, 102, 138, 232

clocks (*continued*)
  pendulum clock invention, 37, 105, 128, 136
  quartz watches and clocks, 48, 96, 102, 103, 140, 148, 246*n*
  sundials, 55, 134, 135
  time base, 102, 103, 140
clock time
  as measure of change, 15, 148, 162
  monasteries and churches as time-keepers, 135, 245*n*
  physics and, 127–43
cognitive load, 61–62
comparison stimulus, 62–63
Computing Tabulating Recording Company, 142
concave–convex illusion, 192
conditioned stimulus, 24, 238*n*
consciousness
  adaptive value, 231
  backward editing in time, 219–20, 222
  binding the past and the future, 215–33
  edited version of reality, 217, 218–20
  neural correlates of consciousness, 220–22
  and neural dynamics, 73–74, 175–76, 249*n*
  presentism and, 216–17
  processing delays between visual and auditory and signals, 218–19, 252–53*n*
  saccades and blinks not perceived, 217
  slices of block universe and, 176, 231
  slowness of, 32, 73–74, 221–22
  spokesperson metaphor, 253–54*n*
  subliminal stimulus and, 221
  temporal window of integration, 218–19
contextual change hypothesis of retrospective timing, 241*n*

contiguity, 23–25
Coordinated Universal Time, 47
Corballis, Michael, 195, 198
credit card use, 205, 208, 252*n*
cross-generational memories, 204–5
cryptodualism, 229
cutaneous rabbit illusion, 220
cyanobacteria, 43–45
cymbals, 191, 218

Darwin, Charles, 8
Davies, Paul, 170
Dehaene, Stanislas, 221
delayed gratification *vs.* immediate gratification, 207–9
Denton, Jeremiah, 85
determinism, 223, 224, 228, 230
Dickens, Charles, 18
discovery of time, 5–8
DNA
  carbon radioisotope ^{14}C in, 129–30
  replication, 45
  transcription, 46
  ultraviolet (UV) damage, 45, 240*n*
Dobzhansky, Theodosius, 231
dopamine, 67, 68
Duncan, David Ewing, 132

early astronomy and telling time, 128
earthquake and tsunami, Japan (2011), 195
ego-moving perspective, 184, 185
Egyptian calendar, 133
Einstein, Albert
  and block universe, 249*n*
  on clock time, 15, 101
  On the Electrodynamics of Moving Bodies, 139, 158
  eternalism, 169–70, 249*n*
  gravity as the curvature of space-time, 150, 168
  illusion of past, present, and future, 157
  meeting with Piaget at Davos, 179
  relative time, 8, 19, 160–61

relativity of subjective time, 190–91
synchronizing distant clocks, 139
time-symmetric equations, 150, 151
see also general relativity; special
relativity
electrocoordination of clocks, 139
electrons
double-slit experiment, 156, 247*n*
resonance frequency, 140
wave function collapse during measurement, 156, 247*n*
elements (chemical), 130
Ellis, George, 147, 228–29, 246*n*
emergent properties in recurrent networks, 123
encephalization quotient, 210
encoding time in changing states of
windows, 112–13
entrainment, 48, 55, 240–41*n*
entropy, 152–55
epilepsy, 54, 225
episodic memory, 196–98, 205, 212,
214
escape from light hypothesis, 45,
239*n*
eternalism
acceptance in physics and philosophy, 13, 149, 168–69, 230
block universe, 12, 13–14, 27,
146–47, 168–69
challenges and failings of, 230–31
compatibility with time travel, 12,
168, 169, 177, 238*n*
defined, 11, 146
"exist," meaning of, 146–47
general relativity, 168
history, 13
moral responsibility and, 229
vs. presentism, 11–14
"real," meaning of, 146–47
spatialization of time, 146
special relativity, 149, 156, 166–68
vs. subjective feeling that time
flows, 147, 174–75
untensed time, 147, 238*n*

Euclid, 6–7
event-specific clocks, 113–16
Everett, Daniel, 203, 214, 251*n*
evolution
adaptive value of consciousness, 231
adaptive value of subjective experiences, 173–75, 193, 231
circadian clocks in cyanobacteria,
43–45
escape from light hypothesis, 45,
239*n*
and multiple clock principle, 32
nervous system attuned to laws of
physics, 173, 178
predictive function of memory,
10–11, 20, 32
primate brain evolution, 144–45,
206
evolving block universe theory, 147,
228
excitatory neurons, 70, *108*, 242*n*
excitatory postsynaptic potential
(EPSP), *28*
eye-blink conditioning, 115

familial advanced sleep-phase disorder, 50
faster-than-light communication, 177,
249*n*
faster-than-light travel impossible,
160
Fee, Michale, 94, 95
Feynman, Richard, 57
fixed-interval procedures, 66–67,
241*n*
flashbulb memories, 71–72
flow of time
adaptive value of subjective sense of,
174–75, 231
block-universe/time-flow paradox,
170–72, 175
ego-moving perspective, 184, 185
eternalism *vs.* subjective feeling that
time flows, 147, 174–75
as illusion, 13, 170–71, 172–73

flow of time (*continued*)
  moments-within-a-moment
    hypothesis, 172, 175, 176, 249*n*
  physics *vs.* neuroscience, 13–14,
    169–72, 216, 231–32
  time-moving perspective, 184
Follini, Stefania, 39
four-dimensional block universe. *see*
  block universe
Franklin, Benjamin, 142
free-running rhythms
  circadian clocks, 37–38, 40, 41, 42,
    46, 50
  hamsters, 40–41, 51
  individual cells, 42
  isolation experiments, 38, 50, 52
free will
  definitions, 222–24
  and determinism, 223, 228
  as a feeling, 224, 228, 253*n*
  and moral responsibility, 228–30,
    254*n*
  neurophysiological studies of, 226–27
  predicting finger movement, 226–
    27
  predicting human decisions, 225–28
  punishment and, 229–30
  and the soul, 223
  time and, 222–24
  and unpredictability, 223
fruit fly (*Drosophila melanogaster*), 45,
  50, 162, 240*n*

Galilei, Galileo
  death, 128
  dynamics, 7
  longitude problem, 138
  pendulum motion, 7, 135–36
  principle of relativity, 158
Gaspard, Enrique, 238*n*
general relativity
  compatibility with time travel, 168,
    169, 177
  equivalence between gravity and
    acceleration, 168

eternalism, 168
  gravitational time dilation, 143, 248*n*
  gravity as warping of spacetime,
    150, 168
  time as part of the fabric of the uni-
    verse, 230
  Wheeler-DeWitt equation, 247*n*,
    254*n*
  see also Einstein, Albert; special rel-
    ativity
geometry, 6–7
Gilbert, Daniel, 199
Goel, Anu, 97
Goodall, Jane, 202
googol, 153
Gopnik, Adam, 253*n*
GPS (Global Positioning System),
  141, 143, 246*n*
Gracián, Baltasar, 3
gravity
  as the curvature of spacetime, 150,
    168
  effect on clocks, 143
  and general theory of relativity, 168
  law of gravity, 199
  quantum gravity, 177
great apes, 127, 201–3, 210, 213
Greene, Brian, 172, 177–78
Gregorian calendar, 133
Gregory (pope), 133
gridiron pendulum, 138
Groundhog Day (movie), 17

H. M. (amnesic patient), 98
hair cells (auditory sensory cells), 85,
  237*n*
half-life, 130–32, 245*n*
Harrison, John, 138–39
hashish, 64, 65
Hawking, Stephen, 223, 238*n*
Heim, Albert, 57–58
hemineglect after a stroke, 189
Hendrix, Jimi, 79, 80
hidden state (neural networks), 110,
  227

hippocampus
  adult neurogenesis, 130
  chain-like patterns of activity, 117
  distance and time neurons in rats, 190
  episodic memory, 198
  location in temporal lobe, 212
  mental time travel, 198
  place cells, 5, 76, 190
Hockenberry, John, 57
holiday paradox, 60
Hume, David, 23, 24, 25, 27, 224
Huxley, Thomas, 224
Huygens, Christiaan, 37, 105, 128, 136, 138, 140
HVC neurons, 93–95, 111
hypermemory hypothesis, 71–72, 75
hypothalamus, 40

immediate gratification vs. delayed gratification, 207–9
Indian Ocean tsunami (2004), 204
Infeld, Leopold, 101, 104
infradian rhythms, 53–55
infraperiod timing, 102–4
inhibitory neurons, 108, 109, 242n
insulin production, 42, 51
interaural time delays, 31, 33, 137
internal clock model, 101–2, 103, 244n
internal dynamics of neural network, 97–98
International Business Machines, 142
International Time Recording Company, 142
Interstellar (movie), 17
intertemporal trade-off: immediate vs. delayed gratification, 207–9
interval discrimination
  by cortical circuits in tissue culture, 96–98
  effect of practice and training, 88–90, 243n
  Goldilocks zone of hundreds of milliseconds, 100, 104

internal dynamics of neural network, 97–98
interval discrimination thresholds, 88–90
intrinsic ability of neural circuits, 96, 98
  by musicians, 89–90
  neuroanatomy of time, 95–99
  pauses between words, 31–32, 79–80
  standard vs. comparison interval task, 87–88
  on a subsecond scale, 80–81, 83, 87–90, 95–100
  voice-onset time of phonemes, 80, 99, 109, 124
  words in motherese, 83
  see also beat; Morse code; patterns and timing in speech
isolation experiments, 37–40, 50, 52
isotopes, defined, 130
Ivry, Richard, 10, 89

James, William, 9, 23, 37, 60, 64, 65
Japan, earthquake and tsunami (2011), 195
jet lag, 48–49, 53
Johnson, Hope, 97
Johnson, Mark, 182
Julian calendar, 133

kappa effect, 186–88, 191, 250n
K.C. (amnesic patient), 198, 212
keeping a beat. see beat
kingfishers, 171, 173, 175, 215
Konopka, Ron, 45–46, 50

Lakoff, George, 182
language and time
  asymmetric relationship between space and time, 185
  children's understanding of time, 181
  "exist" and "real" in eternalism and presentism, 146–47

language and time (*continued*)
  mental time travel and language
    and numbers, 202–4, 214
  motherese, 82–83
  presentist perspective of language,
    147
  prosody, 32, 81, 83, 86
  spatial terms used to talk about
    time, 182–85
  temporal order and interval, 26
  time as most common English
    noun, 3–4, 10, 14
leap seconds, 133, 143
leap years, 133
learning
  classical conditioning, 24–25, 115,
    238n
  synaptic learning rules, 29–31, 121,
    239n
  temporal contiguity, 23–25
Le Guen, Véronique, 39
Leibniz, Gottfried, 7, 138
Lincoln, Abraham, 22
Lockwood, Michael, 238n
logistic equation, 119–20
Long, Michael, 94, 95
Longitude Act, 137
Longitude Prize, 137, 138
longitude problem, 136–37
Looper (movie), 17
Lorentz, Hendrik, 160
Lorentz transforms, 160–61, 247n
Lorenz, Edward, 8
luciferase, 42
luciferin, 42
lunar phases, 53–55

Maass, Wolfgang, 109
Mach, Ernst, 148
Mahabharata (Hindu poem), 18
Mahncke, Henry, 87
Mairan, Jean-Jacques d'Ortous de, 41
marijuana, 64, 65, 66
marine chronometers, 138–39
Mauk, Michael, 112, 113, 115

Maxwell's equations, 150
Meck, Warren, 67
memory storage
  evolutionary function of memory,
    10–11, 20
  increasing focus on time, 10–11
  problem of time, 10–11
  remembering past to predict future,
    10–11, 20, 32
mental timelines, 20, 189, 202, 214
mental time travel
  in animals, 199–202, 213
  anterograde amnesia effects, 197–98
  in the brain, 209–13
  cognitive complexity, 212, 213
  cross-generational memories, 204–5
  death and afterlife, 232–33
  defined, 22, 195
  by early humans, 22–23, 196, 232
  episodic memory, 196–98, 205,
    212, 214
  hippocampus, 198
  language and numbers interdepen-
    dence, 202–4, 214
  prefrontal cortex and, 210–11, 213,
    252n
  semantic memory, 196–98, 204–5,
    212, 214
  and temporal discounting, 207–9,
    211
  temporal lobes and, 212–13
  temporal myopia, 25, 205–9, 233
  see also time travel
Merzenich, Michael, 87
metaillusion hypothesis, 72–75, 78
meter, defined, 143
methamphetamine, 67
Midnight in Paris (movie), 17
Mimosa pudica ("touch-me-not"), 41
Minkowski, Hermann, 19, 167, 168
Moka people in Thailand, 204
moments-within-a-moment hypothe-
    sis, 172, 175, 176, 249n
mondegreens, 80
Montague, Read, 223

moral responsibility and free will, 228–30, 254n
Morse code
  duration of dots, dashes and pauses, 85–86, 87
  Farnsworth timing, 86
  interpretation in brain, 96, 99, 109, 124
  overview, 84–86
  prosody, 86
motherese, 82–83
motion detection by retinal cells, 237n
motor cortex, 96, 226
multiple clock principle, 33, 52–55, 68, 232
multiverse hypothesis, 155
Mumford, Lewis, 142

natural time, 14–15, 16
neural networks
  active state, 110, 112, 227
  and brain dynamics, 116–18
  chaotic activity in, 121
  event-specific clocks, 113–16
  hidden state, 110, 227
  internal dynamics of, 97–98
  interval selectivity based on short-term synaptic plasticity, 107–9
  number of neurons in brain, 27, 109
  population clocks, 111–13, 117–18, 132
  ramping firing rate, 118, 245n
  short-term synaptic plasticity, 110
  state-dependent networks, 109–11
  see also recurrent neural networks
neural trajectories, 76–77, 119, 121, 124
neuroanatomy of time, 95–99
neurogenesis in adults, 129–30, 245n
neuromodulator, 70, 71, 242n
neurons
  action potentials (spikes), 28, 69, 94, 99, 226
  adult neurogenesis, 129–30, 245n
  as biological oscillators, 102, 103
  excitatory neurons, 70, 108, 242n

excitatory postsynaptic potential (EPSP), 28
  image, 28
  inhibitory neurons, 108, 109, 242n
  interval selectivity based on short-term synaptic plasticity, 107–9
  nonlinear output, 119
  number in brain, 27, 109
  postsynaptic neurons, 28–29, 69, 106, 109
  presynaptic neurons, 28–29, 106, 107, 226
  spike-timing-dependent plasticity (STDP), 29–31, 239n
  synaptic connections, 27–29
  time constant of, 69, 242n
neuroscience
  compatibility with block universe, 172–76
  evolution of the problem of time, 8–11
  and flow of time, 13–14, 169–72, 216, 231–32
  neglect of the problems of time, 9–10, 237n
  phrenology and, 8–9
  see also specific topics
Newton, Isaac
  absolute and universal time, 7, 8, 161, 162–64, 248n
  calculus, 7
  deterministic universe, 7
  law of gravitation, 168
  laws of motion, 7–8, 150–51, 153, 173
  simultaneity, 163, 164
  teenager in 1657, 128
  time-symmetric equations, 150–51
now
  Einstein on, 169
  under eternalism, 11, 12, 146–47, 168
  lack of significance in physics, 149, 156
  static nows in Platonia, 171
nuclear proliferation, 129–30
Núñez, Rafael, 182, 183

optic chiasm, 40
orangutans, 201
oscillators
  beat as alignment of biological
    oscillators, 104
  breathing and neural oscillators,
    103
  circadian clocks, 48–49, 102
  infraperiod timing and biological
    oscillators, 102
  in man-made clocks, 48, 101, 102,
    138, 232
  neurons as biological oscillators,
    102, 103
overclocking hypothesis, 69–71, 72,
  74, 242n

parietal cortex and information about
  quantity, 188, 189
Parkinson's disease, 67
Parmenides, 13
Partial Nuclear Test Ban Treaty, 129
Patel, Aniruddh, 91
Paton, Joe, 116
patterns and timing in speech
  in comedy, 81–82
  misheard lyrics in songs, 80
  motherese, 82–83
  other cues in speech, 79–80, 243n
  pauses between words, 31–32,
    79–80
  prosody, 32, 81, 83, 86
  speech rate (speed or tempo), 80, 81
  voice-onset time of phonemes, 80,
    99, 109, 124
  see also interval discrimination;
    Morse code; speech
Pavlov's dog, 24, 25
pendulums
  gridiron pendulum, 138
  pendulum clock accuracy, 136
  pendulum clock invention, 37, 105,
    128, 136
  pendulum motion, 7, 135–36
  pulsilogium, 136

Penrose, Roger, 170, 223
Period gene, 46, 47, 50, 240n
Period protein, 47, 103, 162
phantom-limb syndrome, 73, 242n
phonemes
  defined, 79
  voice-onset time, 80, 99, 109, 124
photoreceptors, 24, 97, 244n
photosynthesis and circadian clocks,
  43–44
phrenology, 8–9
physics
  and clock time, 127–43
  presentism rejection by, 13
  time-symmetric equations, 150–51,
    246n
  see also specific topics
Piaget, Jean, 179–80, 181, 187, 192,
  250n
Pinker, Steven, 26, 176, 179, 249n
Pirahã, 202–4, 214, 251n
pitch as spatial information, 85
place cells in hippocampus, 5, 76, 190
Plato, 8
Platonia (static geometrical universe),
  8, 171
Plautus, 134
Poincaré, Henri, 8
Popper, Karl, 249n
population clocks, 111–13, 117–18,
  132
postsynaptic neurons, 28–29, 69, 106,
  109
precision, 36
predicting finger movement, 226–27
prefrontal cortex, 210–11, 213, 252n
presentism
  acceptance in neuroscience, 13,
    194, 251n
  adaptive value of subjective sense of
    flow of time, 174–75
  and consciousness, 216–17
  defined, 11, 146
  vs. eternalism, 11–14
  "exist," meaning of, 146, 147

humans as innate presentists, 18
incompatibility with time travel,
11–12, 238n
"real," meaning of, 147
rejection by physics and philosophy,
13
tensed time, 147, 238n
presynaptic neurons, 28–29, 106, 107,
226
Price, Huw, 13
principle of relativity, 158, 159
Principles of Neural Science (Kandel et
al.), 10
The Principles of Psychology (James),
9, 37
prior experience and judgment, 193,
251n
probabilistic determinism, 224
prosody, 32, 81, 83, 86
prospective timing, 59–60, 61–62,
65, 67, 127, 130, 132
psychoactive drugs, 64–68, 191
psychology
evolution of the problem of time, 8,
10–11
increasing focus on time, 10–11
memory storage, 10–11
study of children and time, 179,
180–81
pulsilogium, 136
"Purple Haze" (Jimi Hendrix song),
80

qualia, 173
quantum mechanics
entangled particles, 150
particles in superimposed states,
150
and prediction of human behavior,
224
probabilistic determinism, 224
Schrödinger's equation, 150, 155–
56
time as parameter in quantum systems, 230

time-irreversible processes
("arrows") in, 155–56
time-reversible equations, 151, 155
wave function collapse during measurement, 156, 247n
Wheeler-DeWitt equation, 247n,
254n
quartz crystals, 15, 42, 48, 96, 101,
140
quartz watches and clocks, 48, 96,
102, 103, 140, 148, 246n

radiodating, 129–32
ramping firing rate, 118, 245n
rattlesnakes, 24
recurrent neural networks
computer simulations, 119–24
defined, 28, 119
emergent properties in, 123
self-perpetuating patterns in, 121,
122, 123
time-varying motor pattern generation, 121–23
writing the word "chaos," *122,* 123
see also chaos; neural networks
reference stimulus, 62–63
relationalists, 148
resemblance, defined, 23
retrospective timing, 59–60, 61–62,
128–31, 241n
ripples, 104, *105,* 106, 111, 129, 148
Roman calendar, 132
Rosbash, Michael, 46
Rovelli, Carlos, 254n
rubber-hand illusion, 74
running wheels, 34–35, *36,* 41, 117

Sabinianus (pope), 135
saccades, 217
scalar expectancy theory (SET),
244n
Schrödinger's equation, 150, 155–56
scrub jays, 200–201, 213
sea worms, 54–55
second, defined, 140–41, 148

second law of thermodynamics, 152–53, 154–55, 176
selling time, 141–42
semantic memory, 196–98, 204–5, 212, 214
sense of time
  creation by brain, 21–22
  defined, 15
  phrenology and, 8–9
  see also flow of time; subjective time
sense of touch, 5–6
Shakespeare, William, 18
shift workers, 51
short-term synaptic plasticity, 106–9
Shovell, Clowdisley, 137, 141
Shuler, Marshall, 98
Siffre, Michel, 38, 39
simultaneity
  basketball game thought experiment, 157–58
  Newton's absolute simultaneity, *163*, 164
  special relativity and simultaneity, 164, *165*, 166–67, 168–69, 248–49*n*
  subjective relative simultaneity, 191–92, 219
size invariance, 24–25
Slaughterhouse-Five (Vonnegut), 6
slow-motion effect
  causes of, 68–75
  hypermemory hypothesis, 71–72, 75
  in life-threatening situations, 57–58, 68–75, 68–76, 242*n*
  metaillusion hypothesis, 72–75
  overclocking hypothesis, 69–71, 72, 74, 242*n*
  see also subjective time
Smart, Jack, 13
Smolin, Lee, 194
snakes, 24, 182
Snowball, 91
Sobel, Dava, 127

somatosensation, 5–6
Sompolinsky, Haim, 121
songbirds, 92–95, 111, 119
the soul, 223, 229
sound, defined, 222
space, comparison to time, 5–6
spacetime, 150, 166–68
  see also block universe
spatialization of time
  asymmetric relationship between space and time, 185, 187
  concepts of time based on understanding of space, 182, 185
  ego-moving perspective, 184, 185
  under eternalism, 146, 147
  hemineglect after a stroke, 189
  kappa effect, 186–88, 191, 250*n*
  mental timelines, 20, 189, 214
  in neuroscience, 179–94
  in physics, 157–78
  Platonia, 8, 171
  prior experience and judgment, 193, 251*n*
  relative simultaneity and, 166–67, 191–92
  relative time, 190–91
  relativity in physics and neuroscience, 190–93
  spatial terms used to talk about time, 6, 182–85
  STEARC (Spatial-TEmporal Association of Response Codes) effect, 189
  tau effect, 186–87, 250*n*
  time-moving perspective, 184
  see also block universe
spatial maps of vertebrates, 5
spatial summation, 226–27
spatiotemporal patterns
  computer simulations, 111, 121–24
  defined, 104
  population clocks, 117–18
  ripples, 104
  use as timers, 115–16, 117–18, 121

special relativity
  basketball game thought experiment, 157–58, 247n
  block universe, 19, 149, 248–49n
  compatibility with time travel, 177, 249n
  constancy of the speed of light, 159–61, 190
  date published, 19, 158, 160
  eternalism, 149, 156, 166–68
  faster-than-light communication, 177, 249n
  faster-than-light travel impossible, 160
  Lorentz transforms, 160–61, 247n
  loss of simultaneity, 164, 165, 166–67, 168–69, 248–49n
  principle of relativity, 158, 159
  relative time, 8, 19, 160–61
  relativity in physics and neuroscience, 190–93
  space contraction, 160, 248n
  spacetime, 166–68
  time dilation, 160, 161, 248n
  trade-off between space and time, 191, 250n
  train-and-platform thought experiment, 159–61, 162, 163, 164, 165, 166, 247–48n
  twin paradox, 247–47n, 250n
  see also Einstein, Albert; general relativity
speech
  backward editing in time, 219–20, 222
  delay between visual and auditory signals, 218–19
  detecting temporal relationships, 31–32
  motherese, 82–83
  see also language and time; patterns and timing in speech
spike-timing-dependent plasticity (STDP), 29–31, 239n
spokesperson metaphor of consciousness, 253–54n

Star Trek (movies and TV show), 17
state-dependent networks, 111
STEARC (Spatial-TEmporal Association of Response Codes) effect, 189
stopped clock illusion, 63
storage size hypothesis of retrospective timing, 241n
striatum, 116
subjective time
  adaptive value of, 174–75
  ambiguity describing temporal distortions, 64–65
  chronopharmacology, 64–68, 191
  chronostasis, 58, 63
  cognitive load, 61–62
  contextual change hypothesis, 241n
  defined, 15–16
  flow of time as illusion, 13, 170–71, 172–73
  holiday paradox, 60
  in life-threatening situations, 57–58, 68–75, 242n
  overestimation of duration of emotional events, 58
  prospective timing, 59–60, 61–62, 65, 67, 127, 130, 132
  psychoactive drugs and, 64–68, 191
  relativity of, 190–91
  retrospective timing, 59–60, 61–62, 128–31, 241n
  stopped clock illusion, 63
  storage size hypothesis, 241n
  time compression, 61–63, 75–78
  time dilation, 61–63
  see also slow-motion effect
subliminal stimulus, 221
substantia nigra, 67
Suddendorf, Thomas, 195, 196, 198
supplementary motor area, 96, 225
suprachiasmatic nucleus
  entrainment, 48
  lesions, 40, 52, 53
  location, 40, 48

suprachiasmatic nucleus (*continued*)
  master circadian clock, 40–41, 42, 47–48, 51, 56, 103
  transplant experiments, 40–41
supraperiod timing, 102–4, 103
synapses
  excitatory postsynaptic potential (EPSP), *28*
  excitatory synapses, 28, 29
  image, *28*
  inhibitory synapses, 28
  interval selectivity, 107–9
  short-term synaptic plasticity, 106–9, 110
  spike-timing-dependent plasticity (STDP), 29–31, 239*n*
  strength of, 29, 106, 239*n*
  synaptic learning rules, 29–31, 121, 239*n*
synaptic cause and effect, 27–31
synaptic delays, 69
synfire chain, *94*

tau effect, 186–87, 250*n*
telling time
  across time scales, 31–33
  brain as time-telling machine, 21, 32, 232
  event-specific clocks, 113–16
  internal clock model of timing, 101–2, 103, 244*n*
  population clocks, 111–13, 117–18, 132
  ripples as a timer, 104, *105,* 106, 111, 129, 148
  vs. consciously perceiving the passage of time, 32
  see also circadian clocks; interval discrimination
temporal contiguity, 23–25
temporal discounting, 207–9, 211
temporal discounting rates, 207–8, 209
temporal lobes, 212–13, 237*n*
temporal misdirection, 26–27
temporal myopia, 25, 205–9, 233

temporal order, 25–26, 99
temporal summation, 226–27
temporal window of integration, 218–19
tensed time, 147, 238*n*
The Terminator (movie), 17
THC (tetrahydrocannabinol), 65, 66, 67
theory of general relativity, *see* general relativity
theory of special relativity, *see* special relativity
The Time Machine (Wells), 18
time
  challenge of defining, 4, 16, 149, 177, 230
  discovery of, 5–8
  many meanings of, 4
  as most common English noun, 3–4, 10, 14
  and neuroscience, 8–11, 237*n*
  see also specific topics
time base of clocks, 102, 103, 140
time compression
  ability to replay events at high speeds, 76–77
  ability to speed motor actions up or down, 76
  in the brain, 75–78
  in mentally replaying events, 76
  in perspective and retrospective timing, 61–63
time constant of neurons, 69, 242*n*
time dilation
  general relativity, 248*n*
  in isolation experiments, 38–39
  in perspective and retrospective timing, 61–63
  special relativity, 160, 161, 248*n*
time flow, *see* flow of time
Time Machine (Apple backup software), 20
Time Machine (Wells), 18, 238*n*
time-moving perspective, 184
Time's Arrow and Archimedes' Point (Price), 13

time travel
  absence from literature until nine-
    teenth century, 17–18
  "Chronology Protection Conjec-
    ture," 238n
  under eternalism, 12, 168, 169, 177,
    238n
  exotic requirements for, 19, 177
  under general relativity, 168, 169
  incompatibility with presentism,
    11–12, 238n
  in movies, books, and TV, 17, 18,
    238n
  not prohibited by physics, 19,
    238n
  under special relativity, 177, 249n
  see also mental time travel
The Time Traveler's Wife (movie), 17
timing in speech, see patterns and
    timing in speech
toilet flushing, negative feedback
    loops, 46
train-and-platform thought exper-
    iment, 159–61, 162, 163, 164,
    165, 166, 247–48n
  see also special relativity
transcription/translation autoregula-
    tory feedback loop, 46–47, 53,
    102
transistors, 27, 124, 145
tree shrew, 210
tsunamis, 195, 204
Tulving, Endel, 22, 196, 198

Twain, Mark, 206
twin paradox, 247–47n, 250n

ultraviolet (UV) radiation, 45, 240n
unconditioned stimulus, 24
Universal Coordinate Time, 133
universal determinism, 223
untensed time, 147, 238n

van Wassenhove, Virginie, 62
vocal learning hypothesis, 91–92
voice-onset time of phonemes, 80, 99,
    109, 124
Vonnegut, Kurt, 6

water clocks (clepsydra), 134–35
Wearing, Clive, 60–61, 98
Wegner, Daniel, 224
Wells, H. G., 17, 18
Weyl, Herman, 14
Wheeler, John, 144
Wheeler-DeWitt equation, 247n, 254n
wheel running by rats and mice,
    34–35, 36, 41, 117
World Wide Web, 28–29
wormholes, 19, 177
Wright, Beverly, 87

zebra finches, 92–93
zeitgebers, 48
Zeno's arrow paradox, 176
Zimbardo Time Perspective Inven-
    tory, 251–52n